ACTION, ABILITY AND HEALTH

INTERNATIONAL LIBRARY OF ETHICS, LAW, AND THE NEW MEDICINE

VOLUME 1

ACTION, ABILITY AND HEALTH

Essays in the Philosophy of Action and Welfare

by

Lennart Nordenfelt

Linköping University, Sweden

SPRINGER-SCIENCE+BUSINESS MEDIA, B.V.

A C.I.P. Catalogue record for this book is available from the Library of Congress.

ISBN 978-90-481-5412-8 ISBN 978-94-015-9361-8 (eBook)
DOI 10.1007/978-94-015-9361-8

Printed on acid-free paper

TABLE OF CONTENTS

PREFACE

This book is a part of the ongoing enterprise to understand the nature of human health and illness. This enterprise has expanded dramatically during the last decades. A great number of articles, as well as a fair number of monographs, on this topic have been published by renowned international publishers. In this discussion most participants share the idea that health is a partially normative concept, i.e. that health is not a phenomenon which can be wholly characterised in biological (or otherwise descriptive) terms. To ascribe health to a person is *eo ipso*, according to this line of thought, to ascribe a positively evaluated property to this person. Moreover, most debators share the idea that health is a holistic property, belonging to the person as a whole, whereas diseases, injuries and defects are entities (or properties of entities) which can be very limited and and normally affect only a part of the individual.

My own monograph belongs to this tradition. A feature of my position, which is not universally acknowledged in rival theories, however, is my emphasis on the notion of *ability* as a fundament in the theory of health. In my formal characterisation of health I view it as a state of a person which is such that the person has the ability to fulfil his or her vital goals. To put the idea in a simplified way: health is viewed by me as a kind of general ability of a person. This strategy, for which I argue at length in the book, lies behind my decision to base much of my theory of health on modern philosophical action theory, mainly as presented within the Anglo-Saxon analytical school of thought.

A theory of health of the kind sketched here is particularly easy to apprehend and also to apply within those parts of medical sciences and practices where the focus is on human capabilities. A central place in this context has the science and art of rehabilitation. Physicians and nurses in rehabilitative medicine, as well as physical therapists and occupational therapists need a theoretical framework for their practice. They must be able to differentiate between various levels and various kinds of abilities and discriminate between different contexts surrounding, as well as partly determining, these abilities. This book attempts to make progress in providing such a theoretical framework. Thus the book can also function as an introduction to the theory of disability and handicap.

Along with the contemporary interest in health goes a strong concern for clarifying the general goals of medicine. This concern has arisen mainly as a result of two facts; one is the positive development entailing the advancement of medical technology, the other is the limitation of the economic resources available for health care. Expensive medical technology, however efficient and desirable, can be used only to a limited extent. Only a few people, and only on the basis of very specific indications, can be allowed to take advantage of such technology. Such limitations call for principles of prioritisation. Prioritisation in its turn requires the specification of very clear goals with reference to which the priorities should be made.

There have been extensive investigations concerning rationing and prioritisation in almost all Western countries. The United States, Norway and the Netherlands were early in their attempts to formulate guidelines for resource allocation in health care. Sweden has recently published its report and instituted a law for prioritisation in health care (*Priorities in Health Care: Ethics, Economy Implementation*, Final Report by the Swedish Parliamentary Priorities Commission, SOU 1995:5). In all these efforts it has become clear that there is a need for defining the goals of specific medical measures as well as of medicine in general. Thus, some of the governmental reports contain fairly advanced specifications of such goals. These goals are often formulated in terms of life-saving, health and quality of life. Since the border between health and quality of life is particularly difficult to specify much of the intellectual work has been devoted to this distinction. All these attempts call for well-argued conceptual models in the area of health and health care and thus for a development of the genre represented by this work.

The Hastings Center for Ethics in New York in 1993 initiated a broad international research and policy-forming programme to deal with precisely these issues. The Director of Hastings Center, professor Daniel Callahan, gathered representatives from 15 countries (South America was represented by Chile) to consider issues concerning the goals of and priorities in medicine. One of the central issues in these considerations was to characterise the nature of health and its place as a goal of medicine. It is worth noting that the definition of health adopted in the final document of this programme (The Goals of Medicine: Setting New Priorities, *Special Supplement, Hastings Center Report*, November-December 1996) has clear affinities with the notion suggested in this volume. Health is there (p. 9) characterised as "an acceptable absence of significant malady, and consequently by a person's ability to pursue his or her vital goals and to function in ordinary social and work contexts."

A major purpose of my book is to present a systematic conceptual underpinning of a view of health such as the one adopted by the international Goals of Medicine programme.

Much of the material put together in this book has appeared in previous publications of mine. Almost everything is, however, recast and improved. In the following I shall indicate the principal sources for the various chapters.

Chapter 1: Towards an Ontology of Episodes, is a development of parts of section II in my *Events, Actions and Ordinary Language* [64].

Chapter 2: Towards an Analysis of the Concept of Action, is a development of parts of chapter 3 in my *Explanation of Human Actions* [63].

Chapter 3: On Embedded Actions: Accomplishments and Projects, is a development of parts of chapter 3 in my book *On the Nature of Health* [71]. A previous version of chapter 3 has also been published under the title: On the General Concepts of Action and Ability: A Philosophical Introduction to the Theory of Handicaps [72].

Chapter 4: On the Explanation and Determination of Actions, is a development of parts of chapter 3 in my Explanation of Human Actions [63].

Chapter 5: On the Logical Form of Action-Explanations, is published here for the first time.

Chapter 6: On the Logical Form of Interaction, is a development of a paper which appeared in the Swedish volume *Handling och handlingsutrymme*, edited by H. Berglind and G. Aronsson, 1990.

Chapters 7 and 8: Towards a Theory of Ability and Disability, and On the Notion of Health as Ability, are revised versions of analyses in chapter 3 in my *On the Nature of Health* [71]. The passage called "The Two Concepts of Health" is a revision of a part of my Introduction to the thematic issue on health in *Theoretical Medicine, 14*, 1993.

Chapter 9: On the Concepts of Vital Goal and Happiness, is a new text although parts of it draw upon ideas appearing mainly in my *Quality of Life, Health and Happiness* [69]. The appendix on Amartya Sen's theory of welfare is completely new.

Chapter 10: On the Multiplicity of Health Concepts, is published here for the first time.

Chapter 11: On the Technical Notions of Disability and Handicap: The WHO Context, is a revised version of a paper entitled: On the notions of disability and handicap, published in *Scandinavian Journal of Social Welfare, 2*, 1993.

Chapter 12: On the Complexity of Autonomy, has been published under the same title in P-E. Liss and N. Nikku (eds.) *Health Promotion and Prevention: Theoretical and Ethical Aspects* [49].

Chapter 13: A Sketch for a Theory of Health-Enhancement, is based on my booklet *Towards a Theory of Health Promotion* [67].

Chapter 14: On the Role of Compulsion in Mental Illness: The Case of Forensic Psychiatry, is a revision of parts of chapter 4 in my *On Crime, Punishment and Psychiatric Care* [68].

I am grateful to several colleagues and friends for their important advice and support in my endeavour to improve the arguments and clarify the presentations in this book. I am in this respect particularly obliged to Drs Per-Erik Liss, Lars-Göran Johansson and Örjan Widegren at the University of Linköping.

I also wish to thank Mr. Fredrik Svenaeus and Mrs. Sofia Torstensson Kjellström for helping me with the index as well as Mr. Malcolm Forbes for his very substantial help in putting my English into publishable condition. Finally, I am very grateful to Monika Thörnell and Anna Schenell for the tremendous effort they have put into finishing this book.

Linköping, November 1999

Lennart Nordenfelt

INTRODUCTION

General Introduction

It is a salient fact that human happiness, and welfare in general, is highly dependent on human action. A well-functioning society cannot come into being and continue to function, factories cannot be constructed and be kept going, schools and other public services cannot function, without continuous human activity and labour. It is, however, less obvious that there is a deeper relation, a relation of a conceptual kind, between the concept of action and the concepts of welfare, in particular health and happiness.

In the present book I wish to explore and substantiate this idea of a strong conceptual interdependence between the concept of action and some of the central concepts of welfare. As a corollary of this idea I wish to claim that we cannot fully understand the nature of health and happiness if we do not understand the nature of action.

Forming a basis for this presentation are a number of specific studies carried out during the last twenty-five years. These studies have concerned explanations of human actions [63], episodes, including human actions [64], causation [65], the nature of health [71], health promotion [67], mental illness, causation and the law [68] and happiness [69]. Some main results from these analyses are presented here, although mostly in a condensed, and in certain respects revised, form.

My primary intention in putting the analyses together has been to show, stepwise, by beginning with the simplest concepts and ending with the most complex ones, how the conceptual network of human welfare could be constructed. In the second half of the book I draw some conclusions from this construction. I do it by focusing on certain areas central to the science and art of medicine.

The action-theoretic analyses performed in this book play, however, quite different roles for the subsequent discussion on the various notions of welfare and illfare. Let me briefly explain these differences.

a. *The concept of health.* The strongest relation is found in my analysis of the concept of health. I argue that health presupposes the notion of *ability*. Health is here characterised as a person's second-order ability to realise her vital goals. *A fortiori* this analysis has implications for the theory of ability and diasability put forward later in the book.

b. *The concept of happiness*. There is also — although indirectly — a relation between the concepts of action and happiness. In the course of characterising the nature of action and action-explanation, I propose definitions also for a number of mental concepts related to action. Among these are the concepts of intention, want and emotion. Want plays a particularly important role in my subsequent analysis of happiness. I there adopt the old idea that a person's happiness is — per definition — the result of her realising her most important goals in life.

c. *The explanation and determination of health behaviour*. In this book I develop and defend a theory of action-explanation in terms of the agent's intentions and beliefs. I propose — much inspired by von Wright — an intentional explanatory schema which I show has an unexpectedly general scope in characterising action-explanations. Because of this generality the schema can be used also for the investigation of how people can be influenced to act. Such an analysis has a central place in the theory of health education and health promotion insofar as these deal with influencing people's life-styles.

d. *The analysis of autonomy*. Autonomy is a central notion in medical ethics. Respect for the patient's autonomy or for a colleague's autonomy is a corner-stone in most medico-ethical codes today. The term "autonomy" is, however, ambiguous in the following sense: it sometimes connotes a theoretical notion entailing a certain set of capabilities to perform certain acts, whilst at other times it connotes a person's right to have and execute such capabilities. In my own analysis I wish mainly to contribute to the theoretical notion of autonomy and show its intimate relations to the notion of *practical possibility* developed in this book.

e. *The concept of accountability*. Accountability is a basic notion in most legal systems, in particular in criminal law. According to a traditional analysis of the notion of accountability, a person is considered accountable for her deeds only if she could have acted otherwise than she in fact did, i.e. only if she was free to act. Hence, a proper analysis of accountability presupposes a characterisation of freedom of action. In my last chapter I shall propose such an account in the context of a presentation of the basis for modern Swedish penal law.

Most of my analyses are inspired by contemporary works, mainly deriving from the English-speaking community. I am deeply indebted to Ingmar Pörn [78], Anthony Kenny [43] [44], Georg Henrik von Wright [112] [113] [114] and Alvin Goldman [33] in my discussion of action, action-explanation and ability. Ideas of Ingmar Pörn [79] [81] and Caroline Whitbeck [109] have influenced my theory of health, and the analyses by Elizabeth Telfer [99] and James Griffin [35] have been important for the development of my views on happiness.

My project is very much in the spirit of Aristotle. Aristotle has explored, more than perhaps anyone after him, the various notions surrounding human activity and their relation to welfare. He has analysed the nature of ability and action in, for instance, *The Physics* [3] and *The Nicomachean Ethics* [2]. He has commented on the notions of health and medicine in *The Physics* [3]. His famous theory of happiness — *eudaimonia* — is to be found in *The Eudemian Ethics* [4] and *The Nicomachean Ethics* [2].

On one crucial matter I have reached a conclusion which is radically different from Aristotle's. For Aristotle human happiness (*eudaimonia*) *is identical with* a certain kind of activity, viz. virtuous activity. My own notion of happiness lies closer to tnbhe modern common-sense notion of happiness implying a disposition for a particular form of experience. In my analysis there are, however, clear connections between happiness and action, mainly mediated by the notion of a want.

As I have indicated, several of the basic analyses to be presented in this book have appeared before, mostly in publications with very limited circulation. I have, however, here made substantial revisions on several points, partly in response to critical comments. This holds in particular for the analysis of health. Chapter 10, On the multiplic ity of health concepts, for instance, contains entirely new analyses. Completely new is also my appendix on Amartya Sen's theory of welfare included in chapter 9.

Summary of the Contents of the Book

Chapter 1, on episodes, defines some of the most basic concepts to be used in the book. "Episode" is the term used to denote events, processes and states. Thus everything that goes on, happens or lasts for a time is an episode. In the characterisation of these concepts the notion of an object is used as a primitive notion. An object can be in a state, and it can undergo a process as well as a change. In the latter case an event has taken place.

The nature of the object is consciously left undetermined. It is a point for this general characterisation that the object can exist on various ontological levels. It can be microscopic as well as macroscopic. It can be an animate as well as an inanimate object. Moreover, it can be either a natural or a conventional or institutional object.

From this general platform the investigation narrows down to concentrate on episodes in which human beings are not just objects but also *agents*. This means that the chapter focuses on agentive episodes or *intentional human actions.* The concept of human action cuts through the notions previously introduced. There may be agentive states and processes — *activities* — as well as agentive events — *acts.*

Chapter 2 analyses the notion of an intentional human action in a more thoroughgoing way. It characterises first the nature of an intention. An intention is viewed as a mental state which is such that a certain conditional fact holds true. This conditional has two belief-components and a capacity-clause in its antecedent, and an action in its consequent. Roughly, then, an intention can be viewed as a mental state by which an agent is prepared to perform one (or a set of) action(s). Through an intention an agent is prepared to perform all those actions which she considers to be necessary, given a situation, for realising a set of goals.

Chapter 3 attempts to give an overview of the complexity of actions. It categorises actions mainly from the point of view of their complexity. It observes how more complex actions can be built up out of simple ones — *the basic actions*. It also studies the various mechanisms by which the basic actions generate more complex actions.

In chapter 4 the basic analysis of intentional action is put to work in order to clarify explanations of human actions. By explanation is here meant an answer to the question *why* an action has been performed. The chapter starts by describing a paradigm for the explanation of human actions. This paradigm has been deeply inspired by G.H. von Wright's classical study in *Explanation and Understanding* [114]. In this study von Wright develops the old Aristotelian idea of a practical syllogism, viz. a kind of reasoning which has an action as its conclusion. My paradigm for action-explanation is similar to but not identical with the one proposed by von Wright. (For detailed arguments, see Nordenfelt [63].)

In the main bulk of the chapter it is argued that most current forms of action-explanations, proposed in human sciences, in law or in daily affairs, are in actual fact more or less explicit variants of the proposed paradigm. Thus it is argued that explanations of human actions in terms of beliefs, emotions, inclinations and character traits, as well as explanations in terms of external reasons, including external force, can be understood in the light of the intentional paradigm.

This discovery has important implications for the theory of humanistic explanation and also for the theory of determination of actions. Both theories have important fields of application in the health sciences. Questions such as: how do we explain people's health behaviour, and how do we influence people to change their health behaviour? are central in this respect.

In chapter 5 the presented theory of action-explanation is put into the context of the general theory of explanation. I argue that intentional action-explanation is different from the well-known hypothetical deductive explanation of the natural sciences. However, intentional action-explanation is shown to be a version of dispositional explanation and it can be put in the form of a deductive argument. This observation is, I believe, of great relevance for the on-going international debate about the logic of humanistic explanations.

The structural similarity between explanation and determination is illustrated in chapter 6, where the intentional explanatory schema is used for the purpose of analysing various kinds of human interaction. The analysis purports to show that many of the natural kinds of interaction, such as recommendation, advice, persuasion and force, can be separated in terms of influence on different elements in the intentional explanatory schema.

Chapter 7 analyses the notions of ability and disability. These notions are in their turn dependent on the basic notion of *practical possibility*. According to the suggested analysis it is practically possible for an agent *A* to perform an action *F* at time *t*, if and only if *A* succeeds in *F*-ing at *t* if she[1] tries to *F* at *t* or some time immediately preceding *t*. Practical possibility can in turn be divided into two components, viz. ability and opportunity. It is practically possible for *A* to do *F*, if and only if *A* has both the ability and the opportunity to do *F*. *A*'s ability is constituted by her inner, bodily and mental, conditions, and her opportunity is constituted by conditions external to her.

The analysis concentrates on the notion of ability and attempts to trace the various types of conditions needed for the successful performance of actions of different levels of complexity.

The notion of ability thus characterised is a notion which is relative to a specified opportunity. *A* has the ability to do *F*, relative to a particular opportunity *O*. I thereafter go on to interpret the sense of locutions such as: *A* is unable to do her cooking, and *A* is

---

1 My use of feminine pronouns throughout this book is dictated purely by stylistic considerations, nothing more. Such things are especially difficult to manage in a work of this nature, with its many schemata, etc. They are also especially difficult to manage when one is writing in a language not one's own. It goes without saying that no specifically male reference is intended.

in general disabled. Superficially, such locutions do not refer to any specific opportunity in relation to which the disability is identified. How, then, shall they be understood?

These questions bring the discussion to the philosophy of health, where this kind of unspecified ability or disability is paradigmatic. The discussion concludes with the statement that there must be a set of opportunities presupposed in all talk of ability and disability. For the context of health I introduce the notion of *accepted* circumstances. These are the circumstances which are taken for granted by the speaker when ascribing ability or disability to someone. These accepted circumstances can be idosyncratic to the speaker but very often they are identical with a set of circumstances generally taken for granted in a particular culture. In such a case we can talk of *standard circumstances.*

Chapter 8 introduces a further crucial distinction, viz. the one between *first-* and *second-order ability*. The ability which has so far been defined is the first-order kind. By a second-order ability is meant an ability to acquire a first-order ability.

Given this conceptual apparatus the notion of health is defined thus:

> *A* is completely healthy, if and only if *A* is in a bodily and mental state such that *A* has the second-order ability, given accepted circumstances, to realise all her vital goals. (By a vital goal is in turn meant a goal whose realisation is necessary for the subject's long-term minimal happiness.)

The notion of a vital goal requires, then, for its explication an analysis of the notion of happiness, which is the subsequent major task in the treatise (chapter 9). My intuitive idea about happiness can be expressed thus: *A* is completely happy at *t*, if and only if *A* at *t* wants everything in the world to be just as she considers it to be. A more formal and precise way of expressing this thought is the following:

(i) *A* wants at t that (*x1...xn*) shall be the case at *t*
(ii) (*x1*...xn) constitutes the totality of A's wants at *t*
(iii) *A* is convinced at t that (*x1...xn*) is the case.

According to this definition happiness could be said to be an equilibrium between the subject's wants and the world as she finds it to be. A person is furthermore said to be minimally happy if she considers that everything which is of high priority to her has been realised.

In a long appendix to this chapter I relate my analysis of happiness to the much debated theory of welfare presented by Amartya Sen. I show that, in spite of superficial similarities, there are important differences between our views. The appendix contains an argument in favour of my position.

Chapter 10 focuses on the apparent and real multiplicity of concepts of health in the modern literature on health matters. The following questions are highlighted: a. In what sense are there a multitude of concepts of health? b. What should we require of a reasonable concept of health? and c. Is it desirable to have a multiplicity of accepted concepts of health?

The discussion also notices that there are many cases of apparent but not real ambiguity in the health discourse. Important such cases have to do with the relational nature of the notion of health. Since both standard circumstances and vital goals can vary, the former as between cultures and the latter as between individuals, it is clear that two persons with the identical biological make-up can vary as to their health status. But the ascription of different health status to these individuals does not entail the application of different concepts of health.

The presented basic conceptual analyses prepare the ground for some applied studies. The first study (chapter 11) focuses on an analysis performed by the WHO concerning the technical notions of disability and handicap. In this chapter I notice some similarities between on the one hand the approach adopted by the WHO and some other organisations dedicated to this work, and on the other hand my own approach to this problem area. I also attempt, however, to show some weaknesses in the official strategies, and I propose some amendments. For this task I make use of several of the previous analyses.

The second study (chapter 12) discusses the central notion of autonomy, so often used in various medico-ethical contexts. The distinction is made between on the one hand autonomy as a theoretical property of an agent and on the other hand the right to have and execute this property. The main bulk of the chapter is devoted to an analysis of the theoretical property of autonomy. This analysis draws upon the previously introduced notions of practical possibility and ability. A basic sense of autonomy is presented in the following definition:

> *A* is autonomous (in the sense of having the practical possibility) with respect to action *F*, if and only if *A* has both the ability and the opportunity to decide to perform *F*, and has both the ability and the opportunity to execute this decision, given a specified set of circumstances.

In order to approach the fullblown sense of autonomy used in most ethical contexts this characterisation should then be supplemented with a set of requirements concerning the causal story behind a person's actual use of this practical possibility: In order for *A* to be autonomous with respect to *F* she must not be forced, threatened or deceitfully manipulated into deciding to perform *F* or not to perform *F*.

The third applied study (chapter 13) concerns the notions of health enhancement and health promotion. The notion of health enhancement is then taken in the following extremely wide sense.[2]

> A *health-enhancing action* of *A*'s is an action by which *A* has the intention to maintain or improve the health of some individual *B*, where *A* and *B* can be, but need not be, identical persons.

The term health promotion is used to signify that subset of health-enhancing actions which lies outside curative and palliative medical care.

---

2 The notions of health enhancement and health promotion have become more developed in my recent paper [75].

The main task of the chapter is to analyse and categorise various types of health enhancement from the point of view of the causal mechanism employed in the determination of the health-enhancing action. In the case of interactive health enhancement, which is found to be a typical case, the analysis is performed with the help of the explanatory schema introduced in chapter 4. Examples of interactive health enhancement are taken both from curative medical care and from health promotion.

The fourth study (chapter 14) has a very different character and concerns the role of compulsion in mental illness and the conclusions drawn therefrom in the fields of criminal law and forensic psychiatry. I present in this study the so-called positivist (utilitarian) philosophy of criminal law in its present-day Swedish version and I challenge some of its main tenets. I criticise in particular the expulsion of the notion of accountability from the scene of criminal law. The positivists argue that this notion is metaphysical and without any reasonable content. I argue on the contrary that accountability is an indispensable notion. The main steps in my reasoning (which draws upon my analysis of action as well as action explanation above) can be summarised thus:

(i) The notion of a crime as an action presupposes a minimal concept of accountability.
(ii) Every criminal code, including the present Swedish penal code, specifies circumstances which are such that they can exculpate a criminal. If such excusing or legitimating circumstances are present, the malefactor is not held responsible for the crime — in other words he is not held accountable for the crime.
(iii) The traditional notion of accountability emphasises that an accountable criminal could have behaved otherwise than he in fact did. The positivists deny the possibility of free will. Therefore, they conclude, nobody can act otherwise than he in fact does. I challenge this whole reasoning and argue that the locution "could not have acted otherwise" can be given a very plausible interpretation without involving the controversy about free will and determinism. In support of this conclusion I can adduce my analysis of action explanation, and in particular explanation of coerced actions.

PART I

Action Theory

CHAPTER ONE

TOWARDS AN ONTOLOGY OF EPISODES

In our endeavour to describe the course of the world a natural conceptual apparatus has emerged. We use such terms as "change", "process", "state" and "activity". We say that objects undergo changes, we say that people act, or we say that a catastrophe has happened. In fact, the vast majority of sentences produced in a natural language describe, in one way or the other, an aspect of the course of the world.

For my purposes, in particular for the purpose of acquiring a deep understanding of the nature of action and health, I shall attempt to explicate this conceptual apparatus. For the sake of simplicity I shall do this from the point of view of an *object existing in the world*. By an object is here not just meant the small physical object of daily affairs, such as a razor, a saucepan, a book or a bicycle. The object can be a human being, a country or an institution. In general, everything that can be in a state or undergo a change is an object in this theory.

At every instant of its existence an object is involved in a great many *episodes*. The number of episodes is in a way arbitrary, since it is completely dependent on how deeply in detail we wish to scrutinise the object. We may look at it superficially and say that the number of episodes is small, and we may look at it in the utmost detail, for instance by dividing it into parts, and say that the number is very great indeed.

The concept of an *episode* is thus the most general concept to be used in this explication. My next task will be to consider and classify the most important categories of episodes. My main criterion for distinguishing between what I shall here call the elementary episodes is the *change* involved in the episode. The three *elementary episodes* to be discussed are *states*, *events* and *processes*.[3]

States

I shall say that an object x is in a state s at time t, if and only if x at t has a contingent property or stands in a contingent relation of some kind. By a contingent property (relation) I mean a property (relation) which does not belong to (pertain to) its bearer as a logical consequence of the simple fact that the bearer is an object of a particular kind. Examples of contingent properties or relations are: quality (x is red), location (x is in Edinburgh), class-membership (x belongs to the class of animals), and possession (x owns a car). An example of a non-contingent property, which therefore does not constitute a state, is identity (x is identical with itself).

3 The presentation in this chapter is to a great extent based on my *Events, Actions, and Ordinary Language*, [64]. I was there much inspired by some philosophical and linguistic theories about event and action language. My main sources were A. Kenny [43] and Z. Vendler [107]

Most states remain for a period of time. In the following, then, states will normally be interpreted as enduring states. The theory can, however, very well allow for instantaneous states.

Processes

Many objects are also involved in processes. When this is the case the objects undergo a continuous change along some property or relation dimension. The dimension can be distance in space — for the process of movement — or a quality spectrum, like colour, magnitude or shape — for the various processes of quality change.

In order to focus on details, let me choose one of these dimensions, viz. distance. When an object moves, it continuously changes location in space. Typically the locations passed are all different from each other, i.e. the object covers some distance without at any point returning to a previous position. This need not, however, be the case. We say, for instance, that a ball is rolling even if it is moving in a circle and thereby returns to positions that it has previously passed. A necessary prerequisite for *x*'s being in a process of movement is, however, that there are no immediately consecutive times $t\text{-}1$ and t such that x is at exactly the same location at both times. If x is at the same place at two consecutive times we will have to say that x remains at the place and is no longer in a process of movement.

Thus when we claim that an object is in a process of movement at a particular time, this seems by necessity to involve a presupposition about the state of the object at some earlier time. It is to some extent arbitrary to determine what is the minimal number of stages which must have been passed for it to be true to say that there is a process going on. For the following reasons I shall assume that two changes of state of location are necessary and sufficient for the occurrence of a process of movement.

One reason has to do with my definition of event (to be stated below), where an event is taken to be a transition between two states, i.e. taken as involving one change. It seems odd to say that there is a process going on when just one event has occurred. Another reason is that certain more specific process-types — in particular those called repetitive processes below — seem to require for their identification at least two transitions between states.

Although not all processes can be looked at with equally subtle clocks and microscopes as the movement of a ball, I think that the idea about consecutive times can be generalised. As a general characterisation of a process I then propose:

x is in a process at time t, or

$pr(x)_t$, if and only if

there is a property or relation dimension D pertaining to x such that x's position at t with respect to D is distinct from x's position at $t\text{-}1$ with respect to D, and moreover such that x's position at $t\text{-}1$ with respect to D is distinct from x's position at $t\text{-}2$ with respect to D.

(Observe that according to this general definition the positions at *t* and *t-2* may or may not be different. The time points *t-2*, *t-1* and *t* are assumed to be immediately consecutive. This assumption is merely pragmatic and does not presuppose that time is discrete in any absolute sense.)

An important and common special case of processes are the repetitive processes. Repetitive processes are characterised by the fact that they involve an oscillation between a limited number of states. In the simplest and, indeed, the most typical case the number of states is only two. When a lamp is in the process of blinking it is at one instant on and at the next instant off. Analogies to this case but with other types of changes involved include the processes of knocking, hammering, bouncing and laughing.

Processes, like states, can extend over lengthy periods of time. Like states, processes can be maintained. A plumber can keep the water running and the umpire can keep a game going. For this case of a process extending over the minimal period I shall coin the term enduring process.

EVENTS

A common idea in the theory of episodes (see for instance von Wright [112], [113]) is that an event is the coming about of a state. This is a typical case also in my system. Coming into existence, coming to a location, becoming a member, receiving a quality and entering into a relation are certainly all events.

But conversely, it can be added, an event may also be the cessation of a state. Disappearing, leaving a location, giving up a membership, losing a quality and abandoning a relation — these are all events. However, most often the cessation of a state is at the same time the coming about of a new state. Disappearing is becoming non-existent (or becoming invisible), leaving a location is coming to some other location, losing a quality is acquiring some other quality, etc. Observing these symmetries I shall in the following not distinguish the cases of coming to be from the cases of ceasing to be. (From a psychological point of view, of course, there may be reason to distinguish between the two.)

An event may also be the coming about of a process. I shall say that John starts walking, the ball starts rolling and John starts laughing are events which initiate the processes of walking, rolling and laughing. Conversely, of course, the terminations of processes are events. Such a termination is typically the coming about of a state. When John stops walking he enters into a state of rest and when he stops laughing he enters into a state of silence.

More generally then, an event is here taken to signify the transition between states and processes. It may be a transition between two different states, between a state and a process or between a process and a state. If there are immediate transitions between two processes, these should, of course, also be viewed as events.

The most complete way of characterising an event would then involve a description of both an initial stage and an end stage. Since, however, most natural locutions for events only refer to the end stages of the changes in question, I shall allow myself to characterise events along those simpler lines. Let me introduce an operator *Become* with the following interpretation:

x comes into the state s (process pr) at time t, or

Become at t $s(x)$ $(pr(x))$, if and only if

(i) $s(x)_t$ $(pr(x)_t)$

(ii) $not\ (s(x)_{t-1})\ ((not\ (pr(x)_{t-1}))$

At this stage the following question is natural: How do we distinguish a process from a succession of events? According to my characterisations both processes and events involve changes of states.

One answer which covers an important class of cases is to be found already in my definitions. A necessary condition for a series of changes to constitute a process is that the series runs along one single dimension. For instance the series "John takes an exam, John comes home, John goes to bed" cannot qualify as a process. There is no single dimension such as movement that covers all these transitions. Hence, it is just a series of events.

The question could however be pursued with regard to the clearcut process cases. What prevents us from saying that a ball's rolling is a succession of events involving the movement of the ball from one position to another? The answer to this is of course: nothing. The reason for retaining the notion of a process is just that it indicates an interesting, and for many purposes useful, distinction between those series of state transitions which are in some way structured and clearly restricted to one dimension and those which are not.

CAUSATIVE EPISODES

I shall now introduce the distinction between the simple episodes — states, events and processes — and some more complex variants. I shall first introduce *causative episodes*. By a causative episode I mean an episode which in an essential way involves a causal relation. This needs some clarification.

It could be contended that most episodes involve causality in some way or other. For instance, many processes may be thought of as indefinite causal chains. Moreover, according to a popular metaphysics, all episodes have causes.

I shall not here argue about these matters. Even if the metaphysics is granted, there is still room for a distinction between causative and non-causative episodes. This can more easily be put into linguistic terms. A description refers to a causative episode if it actually expresses (and not merely presupposes) a causal relation and mentions both the cause and the effect. Hence, the tree breaks the window (i.e. the tree causes the window to break) is a causative episode, whereas the window breaks is not.

As my example indicates, we here assume that a physical object, for instance a tree, can be the cause of an episode. This is commonly known as the assumption of object causality. This assumption is not unchallenged in the theory of causation. It is plausible to maintain that, in the last analysis, every causal relation holds between episodes or even between just certain kinds of episodes, for instance events.

I shall not dispute this but instead argue along the following lines. (i) My task is to characterise episodes as conceived in ordinary language. In these episodes objects occur as causes. (ii) Even if the doctrine of event causation ultimately turns out to be correct, we may be justified in using locutions designating objects as causes as convenient shorthand expressions.

To signify the causal relation I shall introduce an operator *Cause*. A causative episode can then be characterised as follows:

An object *o* causes the elementary episode *E* at time *t*, or

Cause (o) E_t , if and only if

(i) *o* exists during some period extending at least until *t*
(ii) *E* is the case at *t*
(iii) *o* is a non-redundant part of a sufficient condition for E_t
(iv) there is a transmission of energy from *o* to the object(s) involved in *E*.

(For the sake of simplicity, I here refer to all kinds of elementary episodes by the letter *E*. For the same reason I have simplified the time-reference, which in fact varies between the episode-types.)

By introducing condition (iii) I ally myself to the analysis of causation in terms of necessary and sufficient conditions presented by, among others, John Mackie [50] and Konrad Marc-Wogau [51]. In my own treatment of causation [65] I have discussed some problems surrounding this analysis. By condition (iv) I claim that the conditional analysis of causation cannot be sufficient. There are some kinds of facts, in particular negative facts, for instance the non-existence of an object, which fulfil the conditional requirement but which are rarely referred to as causes. This condition plays a particularly crucial role in my characterisation of omissions below.[4]

I do not view a causative episode as a combination of two episodes. The causing is not per se regarded as an episode. Hence the formula *Cause (o) E* could be given the more accurate reading: there is an episode *E* for which *o* is causally responsible.

As a consequence the causal operator does not change the status of the episode within its scope from the point of view of change. If *E* is an event, then *Cause (o) E* is an event.

Special cases of causative episodes are the *preventive episodes*. By prevention I shall here simply mean the causation of the fact that a particular episode does not occur. In the standard case we talk about prevention in a situation when an episode *is about to occur*. We assume that there is a great likelihood of its occurrence. All or almost all

4 My views on causation will be further clarified in chapter 3. For a detailed account see my [65].

determinants of the episode are present, but in the end some interfering factor causes it not to occur.

Sometimes preventive episodes are also agentive episodes. (See below.) However, they need not be. A natural cause as well as an agent can prevent things from happening. The weather as well as my parents can prevent me from flying to New York.

The episodes prevented can themselves be either causative or non-causative. Consider:

(i) John prevented the rain from ruining the garden
(ii) The storm prevented the window from opening.

How then are the preventions themselves to be analysed? I have already suggested a first approximation. Prevention is causation of the non-occurrence of an episode. When John prevents Peter from going out, he causes it to be the case that Peter does not go out.

This analysis seems uncontroversial so far. A difficulty, however, arises when we attempt to determine the ontological status of these non-occurrences. Are the non-occurrences themselves episodes? In that case is the non-occurrence of an event itself an event?

I shall here propose a short-cut analysis, which aims at doing away with non-occurrences altogether. (The general idea is in line with the procedure to be used for the analysis of omissions below.)

Consider the following examples:

(i) The wind prevented the door from remaining open (the prevented episode is a state)
(ii) The frame prevented the ball from falling down from the cupboard (the prevented episode is an event)
(iii) The rain prevented the game from going on (the prevented episode is a process).

The key to the analysis of these cases is that the sentences (i)-(iii) can be translated in the following way:

(i') The wind caused the door to close
(ii') The frame caused the ball to remain on the cupboard
(iii') The rain caused the game to stop.

(i')-(iii') are all episodes easily identifiable in my framework. (i') is a causative event, (ii') is a causative state and (iii') is a causative event. In general, then, I propose the following way of analysing preventions of non-causative episodes:

(i) The prevention of the continuation of a *state* is the causation of the cessation of the state, which is a causative event.
(ii) The prevention of an *event* is the maintenance of a state, which is a causative state or a causative process.

(iii) The prevention of a *process* is the causation of the cessation of the process, which is a causative event.

The classification made so far can now be exemplified in the following matrix:

	State	*Process*	*Event (s)*	*Event (pr)*
Causative	The pillar supports the bridge	The pump keeps the water running	The wind breaks the window	The germs initiate the deterioration of the organism
Non-causative	The bridge stretches over the river	The water runs	The window breaks	The organism starts deteriorating

THE COMPONENT OF AGENCY: ACTIONS

What is it to act? My strategy in this work — which is indeed a common strategy — is to identify actions with behaviour performed with an intention. This strategy must be explained. I am not claiming that in order to perform an action *F* the agent must intend to do (exactly) *F*. A young boy who kicks a ball so that it breaks the window of the neighbour′s house, normally does not intend to break the window. Still, breaking the window is an action of his. It is an action because the behaviour involved in the breaking is intentional *under some description*. In kicking the ball the boy intends to do something, for instance scoring a goal in a game of football. The sense in which all actions are intentional thus is the following: *F* is an action of *A*'s, if and only if it is true that there is some description *D* of *F* which is such that *A* intends to perform *F* under that description.

An intentional action is in the standard case also a conscious intentional action, but it need not be. My theory is designed to cover also intentional actions of which the agent is unaware or of whose purpose the agent is unaware.

To act, then, is for a human being to *intentionally* make or let something be the case in the world. This means that an action is always put into a cognitive context and is part of a strategy (although often a very simple one) of the agent. (For a closer analysis of intentions, see chapter 2.)

Actions normally involve the agent's body, i.e. bodily movements are often in some way parts of the action. There are important counterinstances to this, however. The most salient examples are *mental actions* and *omissions*.

Actions have *results*. Sometimes the result constitutes the endstate of the action. For instance, the result of the action of closing a door is the fact that the door is closed. Sometimes the result is a state or process existing parallel to the action. Such is the case with the action of playing golf. The result is simply that a game of golf is played.

The result of an action is then identical with that state, event or process created, allowed or maintained by the agent in question. Results of actions must be distinguished from *consequences* of actions. A *result* is tied to an action *by definition* in the way I have just indicated. A consequence is a causal result of the performance of the action. Assume that *A* by closing a door knocks down a vase which happens to be in the way. Assume also that *A* did not intend to knock down the vase. The knocking down of the vase is then a consequence (and not a result) of the closing of the door.[5]

Actions can be subdivided according to the nature of their results. Some actions, like the closing of a door, result in a *state*. Such actions are commonly called *acts*. Some actions involve the maintenance of a process. Such actions, like walking and reading, are commonly called *activities*.

There is an analogy between causative episodes and what I shall here call *agentive episodes* or simply *actions*. In some cases an action indeed entails a causative episode. When John opens the door, part of what he does is that he causes the door to open. Likewise when he walks, part of his action is to cause himself to move. I shall, however, here claim that causation is not conceptually tied to agency, and that there is an important category of actions which does not involve causation. This is the category of *omissions*.

I shall also introduce the notion of a second-order agentive episode. By this I shall mean an action influencing another action, either in a positive enhancing way or in a negative preventive way. These matters will be dealt with in detail in chapter 6 on human interaction.

OMISSIONS

To omit to do something is, in a certain sense, *not to do* it. But not all not-doing qualifies as omission. The answer to the question as to which actions qualify depends at least partly on the general notion of action adopted. In my case I shall say that *A* omits to do *F*, only if *A* believes his abstention from doing *F* is a necessary prerequisite for realising an intention of his.[6]

Two further features are central to my analysis of omissions. (i) An omission to do something is analysed as (intentionally) letting something different occur. (ii) Omissions are analysed as non-causative episodes. I shall briefly comment on these features. (i) The basis for my analysis of omissions into lettings is that there seems to be a clear-cut and general relationship between the two. When I omit to close my door, I let the door remain open. When I omit to keep my door open, I let it close. When I omit to keep the game going, I let the game stop.

---

5 For this use of the terms "result" and "consequence", see G.H. von Wright [114], pp. 88-90.

6 For an alternative, still partly similar, analysis of omissions, see Jari Talja [96]. His definition runs as follows:

An agent *S* omits to perform *a* if
(i) it is not the case that *S* performs *a*,
(ii) *S* has the ability of performing *a*,
(iii) *S* has an opportunity to perform *a*,
(iv) *S* has a reason to perform *a*.

For a recent discussion of the formalisation of omissions, see R. Hilpinen [37].

In general, omission-locutions translate into letting-locutions as follows:

(a) To omit to perform an agentive event is to let a state or process remain the case.
(b) To omit to maintain a state or process is to let an event occur.

This analysis has two advantages. First, it stresses that omissions are actions, pinpointing what is being done rather than what is not being done. Second, it solves the problem of episode classification in a natural way. The omission to cause an event is either an agentive state or an agentive process. The omission to maintain a state or a process is an agentive event. (This theory of classification, of course, presupposes that we, in analogy with the causative episodes, do not view agentive episodes as combinations of two episodes. I shall return to this feature below.)

(ii) When John lets the door close, the door's closing is, in some sense, a result of an action of his. Still, the way in which it is a result differs from the way in which the door's closing is the result of someone's pushing it closed. In the former case we say that the agent has no part in the causal determination of the result. Quite apart from John, we would still have the door closed. Hence I shall say that omissions, although agentive, are not causative episodes.

If we were not to assume that there is agency without causality it would be difficult (if not impossible) for us to distinguish between the following two cases: letting the door remain closed and keeping the door closed. The former is an action of the omission type and the latter is an action which is also causative. Intuitively, the difference between the two is that in the latter case the agent exerts some force — at least now and again — on the door in order to prevent it from opening. There may be a strong wind in the opposite direction which the agent has to counterbalance. This, however, is not the case in the omission alternative. Here nature all the time works in the direction of the agent's intention. No bodily or mental force is required.

To signify the general relation of agency between an agent and an elementary episode, I shall introduce an operator *Do*[7]. An agentive episode can then be characterised as follows:

An agent A sees to it that F at time t, or

$Do\ (A)\ F_t$ if and only if

(i) A exists during some period extending at least until t
(ii) F_t

[7] Parts of the area covered by my analysis here have been carefully formalised and defined within modal logic. Central participants in this discussion have been S. Kanger [42], I. Pörn [78], and R. Hilpinen [37]. I have abstained from entering this formal discussion here. However, my notation is inspired by their work, in particular by Pörn's.

(iii) *A* believes (during some period extending until *t*) F_t to be necessary for the realisation of an intention of his
(iv) *A* can cause or prevent F_t.

The need to include prevention in (iv) seems to be obvious. John does not let a door open (in the omission sense of let) if it happens to be physically impossible for him to prevent this happening. The notions of ability and prevention are left unanalysed at this stage but will be discussed in a later section.

Apart from the case of omission, however, all agentive episodes are in fact also causative episodes. It is then also true that

(v) Some bodily movement of *A*'s is a non-redundant part of a sufficient condition of F_t.

((v) covers the idea about transmission of energy for the agentive case.)
The formula for this case will be: *Do (A) Cause (A)* F_t.

ON THE MULTIPLICITY OF CONDITIONS FOR ACTION

As I have already noted, an action is not the same thing as a bodily movement or an omission of bodily movement. An action must also according to my conception be intentional, i.e. accompanied by an intention.

An analysis of specific action-concepts shows, however, that in many instances there are many further conditions which have to be fulfilled in order for an action to take place. Some actions require for their existence certain results, other actions require tools and certain objects. For instance the act of killing someone presupposes for its realisation not only that a bodily movement has been performed but also that a person has died because of this bodily movement. Hammering on a nail presupposes not only the repeated movement of one's arm up and down, it requires also, per definition, a certain tool, viz. a hammer, which is used during the performance of the action. Or again, the action of driving a car presupposes for its realisation more than just a certain coordinated sequence of foot and arm movements on the part of the agent — it presupposes further the relation of these movements to a certain object, namely a car.

Certain acts require a very complicated set of conditions. One such act is that of getting married. For the marriage to come about, what is normally required — in addition to the bodily movements of the agent in question — is at least the following: existence of a partner, publishing of the banns, consideration of impediments, person to perform the ceremony, witnesses and authorised premises. Thus in various more or less complicated ways an act requires for its existence a set of conditions in the surrounding world. This requirement, moreover, is conceptual and not empirical. The requirement has to do with the definition of the action concept. It has nothing to do with what, as a matter of empirical fact, has to be the case in order for the action to come about[8].

[8] For an excellent classical treatment of actions and their presuppositions, see E. D'Arcy [23].

In chapter 3, "On complex actions", I shall analyse some of the mechanisms behind this complexity of actions. Let it here suffice to say that for many purposes, in particular ethical and legal purposes, it is of the utmost importance how exactly we define a particular action. The circumstantial conditions can here be more important than the behavioural conditions.

A MATRIX OF EPISODES

Thus enriched my matrix of episodes will have the following complexity.

		State	*Process*	*Event (s)*	*Event (pr)*
Agentive					
	Causative	*A* holds the book	*A* keeps the water running	*A* shuts the door	*A* starts the engine
	Non-causative	*A* lets the book lie	*A* lets the water run	*A* lets the door close	*A* lets the engine start
Non-agentive					
	Causative	The pillar supports the bridge	The pump keeps the water running	The wind breaks the window	The germs start the deterioration of the organism
	Non-causative	The bridge stretches over the river	The water runs	The window breaks	The organism starts deteriorating

CHAPTER TWO

TOWARDS AN ANALYSIS OF THE CONCEPT OF ACTION

INTENTION

A crucial concept for the characterisation of action is, as we have seen, the concept of intention. What does it mean, then, to intend to bring about a state of affairs? Is there a particular mental state, an intention, preceding every action? What is the nature of this mental state? Does it, for instance, have to be conscious? And what is the relation between the intention and the subsequent action?

The suggestion to be explored here is that intention is partly a dispositional concept. (I shall return and give a detailed defence of this idea in chapter 5.) Part of the meaning of *"A* intends to bring about *P",* I shall suggest, is that *A* is so disposed that she will bring about *P,* if *A* is capable of and not prevented from doing so.[9]

This provisional characterisation, however, must be supplemented in at least two ways. First, a disposition to act is an intention only under the further assumption that the agent *believes* that she *can* bring about what she is disposed to bring about. As I see it, the notion of intention contains the idea that one cannot intend to bring about *P* unless one believes that one can bring about *P.*

---

9 I am aware that my (partial) characterisation of the concept of intention in terms of disposition for action looks like circular reasoning, given that I have already analysed the concept of action with help of the concept of intention. This circularity can in principle be avoided by replacing the term "action" by the term "behaviour" in this context. I have abstained from doing so since the resulting expressions would be clumsy. A radically different alternative is to construe the concept of action as more basic than the concept of intention. This is the route chosen by B. Rundle [84] where he suggests that action is to be defined in terms of *purposive behaviour* and that intentional actions should form only a subset — albeit the most important subset — of actions. (It appears, however, that Rundle has in mind a slightly narrower concept of intention than the one I suggest.) I abstain from taking an ultimate stand here since my main purpose is not to offer explicit definitions of action and intention. I withhold my preliminary analysis of action in terms of intention, since that is a central feature in some later contexts of my present book, in particular the legal context. I shall also allow myself to use the locution: *A* intends to perform the action *F.*

The philosophy of action and intention has attracted a great deal of interest since I first proposed my analysis. Important contributions are those of R. Tuomela [102], and J.R. Searle [90], C. Taylor [98], C.J. Moya [60], R.M. Mele [57], B. Rundle [84] and R. Audi [5]. With the exception of Tuomela, whom I refer to in chapter 5, C.J. Moya [60] is perhaps the one who has a focus that comes closest to my own. Some of Moya's main claims are the following:

(i) Being intentional is the criterion of agency

(ii) A fully intentional action is a piece of behaviour which follows a commitment on the part of its agent.

(iii) Human agency is an ability to commit oneself to doing things in the future and it is this ability that justifies a sharp distinction between actions and happenings.

(iv) Intentions are not ordinary causes of actions, nor are they factually linked with actions. They prompt actions as standards that the agent commits himself to meeting and to doing so correctly.

Second, if it is true to say, when *A* intends to bring about *P*, that *A* is disposed to bring about *P*, then it must also be true to say that *A* is disposed to bring about all the necessary means for the bringing about of *P*.

Quite frequently it is not a simple affair to bring about an intended state *P*. The agent may have to perform a series of actions, say *F*, *F1* and *F2*, in order to achieve it. Hence, in such a case the bringing about of *P* involves the doing of *F*, *F1* and *F2*. Assume that *A* intends to travel to the country. In order to get to the country she has to pack her suitcase, get down to the station, buy a ticket and finally board the 5 o'clock train. Now, if "*A* intends to travel to the country" is tantamount to "*A* is in a certain way disposed to travel to the country", then it must also be tantamount to "*A* is disposed to pack her suitcase, get down to the station, buy a ticket and board the 5 o'clock train", all of which are actions which are necessary for the realisation of the intended end.

This does not, however, seem to be quite correct. It may not be true that *A* is disposed to perform all those actions which are as a matter of fact necessary for achieving the intended end. *A* may be mistaken about the means. *A* may, for instance, believe that she will get to the country by boarding the 6 o'clock train which as a matter of fact is heading in the opposite direction. And obviously it is *A's belief*, although possibly mistaken, which will determine the choice of action.

Thus, when *A* intends to bring about *P* she is disposed to perform, not those actions which are as a matter of fact necessary for the realisation of *P*, but those actions which she *believes* are necessary for the realisation of *P*. Such a relation of necessity can be of various kinds. *A* can believe the actions to be, for example:

(i) *causally necessary* (e.g. catching a train is a causal requisite for getting to the country in time),
(ii) *conventionally necessary*, or if we wish, causally necessary for conventional reasons (e.g. signing a document is a conventional requisite for getting a passport),
(iii) *logically necessary* (e.g. to run is a logical requisite for winning the running race).

Consider then the following rectified explication:

A intends to bring about *P*, only if
A is in a state which is such that
(i) *A* believes that she can bring about *P*, and
(ii) For all actions *X*, if *A* believes that she cannot bring about *P* unless she does *X*, and if *A* can do *X*, then *A* will do *X*.[10]

---

10 In [63] I suggested the following formalisation of my analysis of intentions:
Int(A, Do (A, P)) = def.
(i) Conv(A,PossDo(A,P))
(ii) (X) ((Conv(A, Unav(Do(A,P) → Do(A,X)) & Poss Do(A,X)) → Do(A,X))).
Conv A means: *A* is convinced that. *Poss* means: it is possible that, and *Unav* means: it is unavoidable that. For the notation here I have been inspired by I. Pörn [77] and [78], as well as by S. Kanger [42].

Observe that this is only a partial characterisation of the concept of intention. It cannot function as an explicit definition, since the concept of action in itself presupposes the concept of intention.

THE IDEA OF NECESSARY MEANS

It could now be asked why, in the analysis of intention, the action required for the intended end need be considered by the agent to be a necessary action. Could it not suffice that it were a sufficient condition or even a part of a sufficient condition for the intended end? Is the following not a current and accepted action-explanation?

(1) *A* intends to bring about *P*.
(2) *A* believes that it is sufficient for her to perform *F* in order to bring about *P*.
—
(3) *A* does *F*.

(From now on I shall allow myself to drop the premise "*A* can do *F*" in such contexts where it is of no particular interest.) This is certainly a frequent and accepted type of explanation, but it is in an obvious way an incomplete one. It is incomplete for the reason that it does not explain why *A* chose just *F* and not some other action, which she also considered to be sufficient for the achievement of *P*. Since *F* is considered to be merely sufficient and not necessary for *P*, then there may very well be another true relevant set of premises of the kind:

(1) *A* intends to bring about *P*.
(2) *A* believes that it is sufficient for her to perform *G* in order to bring about *P*.

If this were an adequate analysis, we should from this be able to conclude that *A* performs *G*, i.e. we should be able to conclude that *A* performs an action which she in actual fact does not perform. An example illustrating this kind of situation is the following:

(3) *A* intends to arrive in Stockholm by 12 p.m.
(4) *A* knows that if she goes on the train that leaves Uppsala at 10 p.m., she will arrive in Stockholm no later than 12 p.m.

On the other hand, the following is also true:

(5) *A* knows that if she goes on a bus that leaves Uppsala at 11 p.m., then she will also arrive in Stockholm no later than 12 p.m.

Assume that *A* in fact chooses the train and that we are requested to present an explanation of this fact, given the above-mentioned data. We see that it would not do to cite just (3) and (4). The schema that the data allows us to construct is the following:

(6) *A* intends to arrive in Stockholm by 12 p.m.
(7) *A* knows that she will not arrive in Stockholm *unless* she goes by train at 10 p.m. *or* by bus at 11 p.m.

(It is here of course taken for granted that *A* believes that there are no further means of getting to Stockholm in time.) What can be explained in this kind of situation, then, is just the disjunctive fact that *A either* goes by train *or* by bus to Stockholm. The choice of means of communication is left unexplained in this account. In general, when the epistemic component of an explanatory argument refers only to a sufficient condition *F* for an intended end *P*, then the conclusion must take the form:

(8) *A* does *F* or one of the other actions which she believes are sufficient for *P*.

The question can now be raised: it there then not a decisive explanation of the fact that *A* takes the train to Stockholm? It may be tempting to believe that there is an alternative explanatory structure of the form:

(9) *A* intends to bring about *P* in the quickest and most comfortable way.
(10) *A* believes that she cannot bring about *P* in the quickest and most comfortable way unless she does *F*.

---

(11) *A* does *F*.

The device chosen here in order to get a complete explanation is to specify the intention in such a way that one and only one course of action turns out to be necessary for the achievement of the end. It may, however, be suggested that there is another way of formulating this schema that would do equally well:

(12) *A* intends to bring about *P*.
(13) *A* believes that if she does *F*, then she will bring about *P* in the quickest and most comfortable way.

The epistemic component here still mentions only a sufficient condition but on the other hand also claims that the sufficient condition in question is the quickest and most comfortable one.

In my opinion this alternative is not adequate. It has the disadvantage that it presupposes that everybody in fact chooses the quickest and most comfortable way to reach an intended end. This need certainly not be the case. The agent may have plenty of time; she may find pleasure in taking alternative routes, etc. Hence, to make the conclusion valid one must specify that it belongs to the intention of the agent that the end should be reached in the quickest and most comfortable way.

There is, however, another important case where it may be claimed that the explanation can be considered to be complete but where the epistemic clause does not refer to a necessary, but only to a sufficient condition. This is the case where the agent believes a particular action to be sufficient for a certain end, where she does not believe it to be necessary, but where on the other hand she does not know of any alternative action sufficient for achieving the end.

Assume that our traveller to Stockholm knows about the fact that there is a train connection from Uppsala to Stockholm. Hence, she is aware of one sufficient means of getting to Stockholm. Assume also, however, that she does not know or believe that there are buses from Uppsala to Stockholm, i.e. that she does not know of any alternative sufficient means of getting to Stockholm. Assume, finally, that she does not believe that the train connection is the only connection to Stockholm. She allows for the possibility that there are other means of communication. Schematically:

(1) *A* intends to arrive in Stockholm by 12 p.m.
(2) *A* believes that if she goes by train at 10 p.m., then she will arrive in Stockholm not later than 12 p.m. *and* it is not the case that there is another action *F* such that *A* believes that if she does *F*, then she will arrive in Stockholm not later than 12 p.m.

This schema seems satisfactory because it is enough to prevent the possibility of there being another set of premises including "*A* intends to arrive in Stockholm by 12 p.m." which will have another conclusion. It would then seem to be too strong a requirement to say that the action to be explained must be considered by the agent to be a *necessary* action for achieving the intended end.

A more penetrating analysis may, however, show that even this case ultimately collapses into the requirement of a necessary action. (1) and (2) constitute a satisfactory explanation of the fact the *A* goes by train only under an important assumption. What this assumption amounts to is that *A* must believe that it is at least improbable that there is an alternative sufficient means for getting to Stockholm which is preferable to the sufficient condition that she knows. If *A* did not believe this, then we would expect her to (try to) find out whether there is some more advantageous alternative than the one that she knows of. Hence when we accept (1) and (2) as a satisfactory explanation of the fact that *A* chooses the train, we must have assumed that *A* believes it to be improbable that there is such an alternative, and as a result of this *decides* not to find out whether this is so or not. And by deciding this, *A* herself in fact creates a situation in which catching the train becomes a necessary action to perform in order to achieve the intended end. Schematically:

(3) *A* intends to bring about *P* in a reasonably convenient way.
(4) *A* believes that if she does *F*, then she will bring about *P* in a reasonably convenient way.
(5) *A* decides not to find out whether there are other actions *X*, such that if *A* does *X*, then she will bring about *P* in a reasonably convenient way.

From (4) and (5) we get:

(6) *A* believes that, under the circumstances (created by herself), she cannot bring about *P* unless she does *F*.

---

(7) *A* does *F*.

In conclusion: in my dispositional analysis of the concept of intention, the consequent of the dispositional conditional must refer to such actions as the agent considers necessary for the realisation of her goal. Observe that this need not entail any serious restriction on the application of the concept. The agent need not consider a particular action to be necessary for her intended end in any absolute sense. It may well be that the agent has no time to explore whether there is more than one action which is sufficient for the intended result. The agent may know that there is a train which will take her to Stockholm. She has no time or no inclination to find out whether there are other suitable means of communication. Given this premise pertaining to the agent's decision situation, she must consider taking the train to be a necessary means to reach Stockholm in time. Thus, taking the train is considered by the agent to be necessary given the circumstances which are partly created by the agent herself.

It could now, however, be asked: must we not add that the action should be believed to be both necessary and sufficient? This does not seem to be required. Consider the following unproblematic case. *A* intends to visit a friend of hers, who lives some blocks from Stockholm's Central Station. One action necessary for being able to visit her friend, then, is that she travels to Stockholm Central Station either by bus or train. The following explanation seems to work correctly:

(1) *A* intends to visit her friend in Stockholm.
(2) *A* believes that she will not be able to visit her friend unless she travels to Stockholm Central Station.

---

(3) *A* travels to Stockholm Central Station.

A knows, however, at the same time, that travelling to Stockholm Central Station is not a sufficient condition for setting herself in the position to visit her friend. There is a further action to be performed before she has achieved this. She must also walk a couple of blocks to reach her friend's house. Hence there is also the following explanation to be construed:

(4) *A* intends to visit her friend in Stockholm
(5) *A* believes that she will not be able to visit her friend unless she walks from the station to her friend's house.

---

(6) *A* walks from the station to her friend's house.

The explanandum of a particular action-explanation *need not,* then, be believed by the agent to be (besides being a necessary condition) a *sufficient* means for the realisation of the intended end. There is, however, one matter of importance to be noted in this context. Our analysis of the notion of intention requires that when *A* intends to bring about *P,* she also believes that she *can* bring about *P*. Hence, whenever *A* intends to bring about *P* and believes *F* to be a necessary but not sufficient condition for *P*, it must be true of *A* that *A* believes that there is a set of actions *X,* performable by *A,* such that *A*'s doing *F* in conjunction with *A*'s doing *X* is *jointly sufficient* for P.

INTENTIONS VERSUS DECISIONS

How do intentions come about and cease to exist? According to a common and very reasonable view, intentions are such states of an agent as are the result of a certain kind of mental actions, viz. *decisions*. On this view, to decide is tantamount to forming an intention. An even stronger contention is to say that intentions can occur only as a result of decisions.

I shall in the following adopt the weaker view of the connection between decisions and intentions. I shall say that decisions are the typical causes of intentions and that "to decide to bring about *P*" in fact means: to form the intention to bring about *P*. I shall, however, also allow for the possibility that intentions may arise "in the course of nature". Analogously, intentions may cease to exist either as a result of an action on the part of the agent (the intention can be withdrawn) or as a result of the course of nature (the intention is forgotten or otherwise dropped).

There is an important theoretical reason for allowing that intentions can come into being without any decision. If decisions are actions, then they must be intentional. But if intentions can come into being only through decisions, then we end up an in an infinite regress. Some initial intention must come into being without a previous decision.

ON WANTS AND THE RELATION BETWEEN WANTS AND INTENTIONS

A further important notion closely linked to intentions (and which will play a central role at a later stage of this book) is the notion of want.

Both intentions and wants may be regarded as species of the more general concept of will. Wants, like intentions, are forward-looking and directed to the realisation of a future state of affairs. Like intentions, wants are dissolved (or changed to something different) when the wanted state of affairs is realised. It is also clear that wants, like intentions, typically end up in action.

To some theorists observations like these have been sufficient to warrant treating wants, at least for all philosophical purposes, as being on a par with intentions. To me, however, there are good reasons for making a differentiation. Although ordinary language does not give very clear indications, there seem to be some important distinctions to be made between the two concepts. Let me here confine myself to observing the following:

(i) Intentions are more intimately linked to actions than wants are.
(ii) Wants can be conflicting in cases where the corresponding intentions cannot. It is possible at the same time both to want *P* and to want *non-P*. It is, however, contradictory to at the same time intend *P* and intend *non-P*.
(iii) Wants must be distinct from intentions, since wants typically result in intentions.

In the following I shall propose a (partial) analysis of wants which takes these observations into account, but which also stresses the similarities between wants and intentions.

In analogy with intentions I shall suggest that wants are a kind of dispositions. I shall not, however, say that they are dispositions to act (in general). They are instead dispositions to the formation of intentions, either dispositions to decide (i.e. intentionally forming an intention) or dispositions of a non-agentive (natural) kind where the consequent is an intention.

To want to bring about a state of affairs *P,* then, is, I shall say, to be disposed to intend to bring about *P.* What then is the more specific nature of this disposition? Consider below the conditions under which a particular agent's want to perform *F* may result in a decision to perform *F.*

A first obvious condition is that *A* knows or believes that she is capable of and not prevented from doing *F* in the circumstances in which she is placed. *A* may very well want to go to the moon but still not decide to do so, since she knows that she is not capable of getting there.

To this we have to add a further condition. *A*'s want to do *F* will result in an intention to do *F* only under the proviso that there is no other want of hers which for its realisation requires her not doing *F* and which overrides her want to do *F.* Assume that a little boy wants to buy some sweets in the grocery-shop. At the same time, however, he feels that he should not buy them, since his mother has forbidden him to eat sweets. Hence, he also has what might be called a duty-want to abstain from buying the sweets. He knows that he is capable both of buying the sweets and of abstaining from buying them. Still, he cannot decide to do both; he must finally decide to do either one or the other of the two. One of the two wants will, as we say, *override* the other. Hence, with wants there is a possibility of subjective conflicts which has to be taken account of in the explication of the notion. Consider:

> *A* wants to bring about *P,* only if
> *A* is in a state which is such that if *A* believes that she can bring about *P* and if there is no overriding factor (such as a conflicting want), then *A* will (intentionally or not) form the intention to bring about *P.*

A conclusion that can be drawn from this interpretation is that when wants are cited as explanantia for actions other than decisions, they are not the most proximate explanantia. They explain indirectly through the mediation of an intention. When there is an explanation of an action in terms of a want, there must also, according to this view, be a more proximate explanation of the same action in terms of an intention. In retrospect, all materialised wanted actions were also intended actions.

CHAPTER THREE

ON EMBEDDED ACTIONS: ACCOMPLISHMENTS AND PROJECTS

INTRODUCTION

In chapter 1 I distinguished between elementary episodes and more composite episodes, viz. causative and agentive episodes. In the present chapter I shall introduce a different kind of complexity and show some mechanisms for creating such complexity.

We sometimes say that a complex episode *E* contains another episode *E1* or that *E1* is *embedded* in *E*. A very common linguistic strategy for expressing this relation is to use the preposition "by", as in the following locutions: object *o* causes an episode *E* by causing another episode *E1,* or, the agent *A* performed action *F* by the performance of *F1*. Let us call the former episode the *main episode* and the latter the *collateral episode.* The main episode can have several collateral episodes.

Examples:

The wind made the house uninhabitable by blowing off the roof.
John started the machine by pressing the button.

THE CONCEPTS OF BASIC ACTION AND ACTION GENERATION

For the purposes of this book I shall in the following exclusively consider the agentive case and thereby introduce a further set of concepts.[11] The first concepts to be introduced are *basic action* and *action generation*. In order to understand these notions let us consider some examples of action embedment. *A* travels to New York by taking a boat. *A* greets a friend by waving her hand. Consider now how the relation can be extended:

The driver started her car
by turning the key
by twisting her hand.

The teacher prevented a quarrel in her classroom
by warning the pupils
by reminding them of the rules of the school
by talking in English
by using her speech-organs.

11 I owe the idea of action generation to A.Goldman [33]. The general phenomenon has of course been observed by many action theorists. Cf. E.D'Arcy [23] and D.S. Shwayder [94]. D. Davidson [22], pp. 53-55, uses the term "the accordion effect" for the phenomenon of action-generation.

The question can now be put: is this series of embedments indefinite? Can one always go on finding further actions by which one performs other actions?[12] The answer to this is no. There is always some action which initiates the chain; this is the so-called basic action. A basic action is an action which is not performed by the performance of some other action. In the standard case the basic action involves just the (intentional) movement of a part of the body. (Under special circumstances the basic action can also be constituted by omitting to move a part of one's body.)

Why can't we say that there are actions behind the basic action? Are there not causes of basic actions in their turn? The answer is: there are certainly causes of basic actions, for instance neurophysiological ones. But these causes are *not* actions; they are not events which are intended by the agents. The subject does not intend that certain neurophysiological events shall take place. Hence the *basic actions are the first intended episodes* in the chain connected by the by-doing relation.

The chains of actions indicated above may be said to be *generated* by the basic action. The basic action of twisting one's hand generates the action of turning the key, which in turn generates the action of starting the car. At the same time there is an important sense in which the basic action is identical with the generated actions. The agent does not first twist her hand and then turn the key. The fact is rather that she twists her hand and, given the circumstances, this action is at the same time the action of turning the key, as well as the action of starting the car.

Is there, then, a last member of such a chain of generated actions at the other end of the chain? Now, this question cannot be answered *a priori*. It seems one cannot draw a limit for conceptual reasons. It depends in the single case on how much is included in the *agent's intention*. If, in the second example, preventing the quarrel actually was the final purpose of the teacher (which may be a plausible hypothesis), then this is the last element of the generated chain of actions in question.

What, then, is the nature of the process of generation itself? What does the expression "by doing" signify? This has been debated and quite thoroughly analysed in recent action theory. One of the most influential discussions is in Goldman [33]. He distinguishes between four kinds of "level generation", as he calls it: (a) causal, (b) conventional, (c) simple, and (d) augmentative.[13]

---

12 In this chapter I use the loucution "performing one action by performing another action". It seems, then, that I take a stand in the controversy concerning whether there is one or an indefinite number of actions involved in an actioin chain. Is "twisting one's hand" the same actions as "starting the car" or are they different actions? My way of expressing myself does not, however, mirror any particular ontological commitment. In fact I avoid addressing that problem here. My inclination on this issue is towards the position advanced by D. Davidson [22], entailing that we are dealing with different descriptions of the same action rather than different actions.

13 The theory of activity presented by the Russian psychologist Aleksei Leontiev, for instance in [47], has features which resemble the theory proposed here. However, there are also important differences: Leontiev distinguishes between the notions of *activity, action* and *operation*. The relation between these categories is similar to the relation of embedment analysed here. An activity in a sense consists of a series of actions and an action is performed by the execution of a set of operations. However, the three categories are ontologically separate in Leontiev's structure. The concept of activity is the most general concept. An activity is necessarily motivated. There can be no activity without a motive. Actions are the basic components of an activity and actions have conscious goals. Apart from its intentional aspect an

The first two kinds are the most important and exhaust the vast majority of existing types of action generation. Let us consider their nature in more detail.

First, *causal* generation. Here there is a causal relation, but it does *not* obtain between the actions in the chain, but instead between an action on a lower level in the chain and the *endstate* of an action on a higher level. So in the case of starting the car by turning the key around, it is improper to say that the turning of the key is a cause of the starting; the turning of the key is a cause of the fact that the engine starts, i.e. the endstate of the action of starting the car.

In short: when it is true to say that a person *F*-s by *G-ing,* and the generation involved is causal, then the endstate of *F*-ing is caused by the *G*-ing.

In the case of *conventional* generation, the generation is effected by conventional stipulation. There is a socially determined rule which says that when a certain action occurs in a particular context it should *count* as some other action. Examples: lifting one's hat when meeting another person *counts* as greeting that person; making certain laryngal noises in appropriate sequences can count as performing an action of speaking; signing a document in the presence of a bank-official can count as numerous actions, for example making an agreement, receiving a loan or buying a house; the moving of a piece of wood can, given proper circumstances, count as a move in a game of chess.

In these cases the relation (of convention) can be said to hold between the actions themselves: a *G*-ing in a certain context *is* an *F*-ing. Another way of putting it is to say that *G*-ing, given proper circumstances, generates the conventional result, which is the endstate of *F*-ing.

Now, given the concepts of *basic action* and *generated action* we see that human beings can do things whose results and endstates vary from being very close to the person to being extremely distant.

Although it is true that we say that a person does or performs all the elements contained in a chain of generated actions, there is an interesting difference between the levels generated. The levels which are closest to the agent are more under her control than the levels further away. Twisting one's hand is more under one's control than starting one's car. There are many things that can go wrong in the causal sequence from the basic action to the endstate, whereas only a limited number of things can prevent the basic action from being performed. Hence, very much of what we do that entails actions beyond basic actions is *dependent upon the course of the world outside ourselves.*

When we say that we intend to open a window, we can never know with certainty that we shall actually succeed in doing so. Our success is dependent upon certain states of affairs in the world (such as our being able to manipulate a standard locking mechanism on the window) which we normally correctly believe to obtain. Thus when we claim that there is a causal sequence from the basic action to the endstates of the ensuing actions in the chain, this does not mean that the basic action is a sufficient cause of these endstates. There must be other factors in the external world which contribute to

action has an operational aspect (how it is done), which is defined not by the goal itself but by the objective circumstances under which the action is carried out. An *operation*, then, is the procedure by which an action is realised. But since an operation in itself does not have a goal it is not a kind of action. Thus operations cannot be identified with the basic actions of my own system.

the coming about of the endstates. That this is so does not entail any peculiarity in our use of the term "cause" here. In practically all uses of the term "cause", be it in ordinary language or science, it signifies only part of a sufficient condition.[14]

There is an analogous situation in the case of conventional generation. The basic action itself is rarely sufficient to be conventionally counted as some other action, but requires certain external circumstances in order to be so.

Consider now some further features of action-generation. We say that the basic action contributes causally or conventionally to a certain result. The question now is: is the basic action also *necessary* for this result? The answer to this is that it varies.

In the case of ordinary causation a cause (including the case where the cause is a basic action) is rarely strictly necessary for its effect. There are often alternative ways of achieving a certain end. In order to prevent a row in a classroom the teacher could use other means. She could, for instance, call for the principal of the school and let this person do the job. Or she could intervene by expelling the most difficult boys from the classroom.

In the case of human action it is not only important, however, to determine whether a basic action is as a matter of fact necessary for its effect. In many contexts, for instance in explanation, it is of greater importance to know whether the agent herself *considers* an action necessary for a certain result. If she does, *for her* there is only one way to bring about that result.

In the case of conventional generation, a relation of necessity between a basic action and its effects need not be a rarity. The reason is that such necessity can always be *stipulated.* We can always stipulate that the *only* way in which *A* can make a will is by signing a particular kind of document. We can stipulate that the only way that Sweden can declare war is by the prime minister's putting his signature to a government decision.

THE CONCEPTS OF PROJECT, ACTION-CHAIN, ACTION-SEQUENCE AND OPPORTUNITY FOR ACTION

We have seen how a basic action can generate actions on higher levels. In the following such generated actions will be called *accomplishments.* The chain of actions related by causal or conventional generation, from the basic to the final accomplishment, will be called an *action-chain.*

Many of our actions in ordinary life turn out to be accomplishments within this theoretical framework. Further reflection shows, however, that most of them entail more than just a basic action plus the course of nature or convention. This simple structure presupposes that there is an *opportunity* for action. Consider the case of the driver. A trivial but indeed necessary precondition for her starting the car is that she puts herself in the right position, i.e. that she sits down on the driver's seat. The accomplishment of starting the car thus in practice presupposes (but does not, strictly speak-

14 For more elaborate discussions on the sufficiency and necessity of causation, see J.L. Mackie [50].

ing, include) the actions of entering the car and sitting down in the right place. These three actions form what I shall call a *sequence of actions.*

Sequences of actions can have a purpose or lack a purpose. (Observe, of course, that all the separate actions, according to the basic definition, are intended. This does not entail that the composition of actions into a sequence need be intended.) As an example of a sequence which lacks a purpose we can consider the following: *A* scratches her nose, *A* answers the telephone, *A* walks towards the door. There is no particular purpose common to these actions and thereby connecting them. They just happen to follow one after another.

A paradigm example of a sequence which has a common purpose is the one that we have previously considered, viz. the sequence of actions leading to the starting of a car. Here the agent performs every element of the sequence with the overall purpose of having the engine started. An action sequence which has a purpose will in the following be called a *project.*[15]

Projects can differ in their nature, partly depending on whether or not their elements are strictly interdependent. Some activities are strictly ordered in the following way: accomplishment *a* provides the opportunity for accomplishment *b;* and *b,* in turn, provides the opportunity for *c.* A simple example illustrating this case is the activity of building a wall of bricks: laying the first row of bricks provides the opportunity for laying the second.

Other projects are not ordered in this strict sense of one providing the opportunity for the other. A child's playing with balls may be a loose project of the latter kind. Her playing with one ball does not normally provide an opportunity for her playing with a second.

In one respect there is an important analogy between accomplishments and projects. Just as many accomplishments can be performed by different means, so can many projects. In other words, an accomplishment is not, in general, defined by a particular action-chain, nor is a project in general defined by a particular action-sequence.

Consider now a little more closely the notion of an *opportunity.* I have said that all accomplishments require opportunities. This truth is a consequence of the fact that basic actions are not sufficient conditions for the endstates of the actions they generate. The agent may have to be at a certain place, at a certain time, and so on, and the causally relevant conditions in the world may have to be normal.

The idea of an opportunity is, however, equally important in the case of conventional generation. Here it may also be easier to describe the opportunity exactly. This has to do with the fact that conventional generation requires for its existence a rule (often explicitly formulated) which the agent must have learnt in order for her action to result in a conventional accomplishment. In the rules of a game, for instance, it is explicitly stated under what circumstances a particular move can be made (i.e. what constitutes an opportunity for the move in question). The civil law contains an enormous catalogue of opportunity descriptions, i.e. descriptions of situations in which various administrative and legal actions can be performed.

---

15 The term "project" is here used for the first time in this sense. I find it more suitable than the term "activity" which had the same definition in *On the Nature of Health* [71].

CHAPTER FOUR

ON THE EXPLANATION AND DETERMINATION OF ACTIONS

INTRODUCTION

So far I have attempted to characterise the notion of action from various angles. I shall now turn to the clarification of explanation of actions. By an explanation of actions I shall here mean an answer to the type-question: *Why does A do F?* For instance, why did Peter enter the bookshop?

In what ways are actions normally explained? If one looks at ordinary language, one will find a varied collection of locutions used for the purpose of explaining actions. The following list may perhaps contain the most important types:

I. Explanations of actions by reference to a mental property of the agent:

- (i) by reference to the agent's intentions or aims ("A went into the bookshop because she intended to buy a book" or simply "went into the bookshop in order to buy a book");
- (ii) by reference to the agent's beliefs ("A closed the window because she believed that it was going to rain");
- (iii) by reference to the agent's wants ("A killed her neighbour because she wanted to marry the latter's husband");
- (iv) by reference to the agent's emotions ("A hit her son because she was angry with him");
- (v) by reference to the agent's attitudes or character traits ("A worked hard because she was ambitious");
- (vi) by reference to the agent's sensations ("A walked up and down because she was in great pain");
- (vii) by reference to the agent's perceptions ("A called for the ambulance because she had observed a car-crash").

II. Explanations of actions by reference to a fact external to the agent:

- (viii) by reference to an *external episode* ("*A* jumped on to the pavement because there was a car coming towards her");
- (ix) by reference to *compulsion* ("*A* handed over the bank's money because a robber forced her to");
- (x) by reference to a *norm* ("*A* thanked the hostess for the dinner, because that is the norm in Swedish society").

This is just a provisional and superficial classification indicated by ordinary locutions. An examination of the listed examples will enable me to make another and more enlightening systematisation.

Let me just insert a note concerning the terms "motive" and "reason". These are terms very often used in action explanations. In the following I shall completely avoid the term "motive". The main reason for this is that ordinary language seems to permit us to refer to practically any explanatory factor of an action as its motive. Intentions, wants, emotions and attitudes can all in most contexts be referred to as motives when they occur in explanatory contexts. Sometimes external facts are also called motives.

The same observation applies to a great extent to "reason". There is, however, a fairly clear, more narrow use of "reason", where reasons are contrasted with intentions and wants and taken to refer (primarily) to facts external to the agents. I shall adopt this more precise concept of reason and use the term "rational explanation" for explanations in terms of such facts.

THE SCHEMA FOR INTENTIONAL EXPLANATION

It will be argued here that all the mentioned kinds of action explanation have a common ultimate analysis. The core notion in this analysis is the concept of *intention*. It can be shown that all the mentioned explanation types are elliptic variants of — as I shall call it — a complete intentional explanation, using a *complete intentional explanans*.

The structure of this explanation is the following:

(i) *A* intends to bring about P
(ii) *A* believes that she is in situation *S*
(iii) *A* believes that in *S* she will not bring about *P* unless she does *F*
(iv) *A* can do *F* in the situation in which she is
____
(v) *A* sets herself to do *F*

(i) to (iv) constitute together what I shall call a *complete intentional explanans* of (v). What I mean by this is that the fact *A* sets herself to do *F* is in a way completely explained by referring to her intention, her two kinds of beliefs and her ability. This contention concerning completeness must not be misunderstood. It does not rule out other, more distant explanations of the same action, in terms of, for instance, the agent's ideology or some environmental influence. What is considered complete here is only the most proximate explanans. The intentional explanans contains those factors which in time are the most proximate explanantia of the action to be explained.[16]

---

[16] This schema is an elaboration of one given in von Wright [114]. Observe that this is not the final and most careful formulation of the schema. Such a formulation must also contain timeclauses. We would have to give requirements concerning the maintenance of the intentions and beliefs over the period of time lastgoing to the moment of the execution of action. For more exact formulation see my presentation in [63], pp. 71-72.

My point that many common-sense explanations can be seen as incomplete variants of a full intentional explanation opens up an interesting problem area which is not addressed by me here. One may ask why,

Before applying this model of action explanation to the analysis of a variety of explanations in ordinary language and human science I shall attempt to answer a criticism which is sometimes raised against it. This criticism concerns the scope of the model; it claims that the model has too strong presuppositions concerning the nature of human action to have any practical implications; in particular it claims that the model presupposes that the agent whose action is to be explained is an *ideally rational person.*

This criticism misunderstands what the analysis of the intentional explanans entails. The explanans does *not* present the thinking of an ideally rational agent. Observe that there is nothing presupposed about the rationality or the quality of the beliefs contained in the second and third premises of the explanans. The beliefs may be false and they may be the consequence of incoherent thinking.

Similarly, the intention contained in the explanans may be counter-productive with regard to the agent's ultimate purposes. She may, for instance, through incoherent reasoning have come to the conclusion that a particular state of affairs is an appropriate means for realising the ultimate purpose despite its actually not being so.

Thus the explanans does not express any rationality on the part of the agent. What the explanans entails is the following: Given that premises of types 1-4 are true, then for logical reasons a conclusion of type 5 must be true. In other words, given that an agent has an intention and a certain set of beliefs, and given that she is capable and unprevented, then she must act in a certain direction. If she *does not* act, then there must be something wrong with one or more of the premises. Either it cannot be true that she *really* intends, or it cannot be true that she *really* believes what she is claimed to believe, or, finally, that she is not capable or unprevented.[17]

By way of summary: my model for the intentional explanans is not exclusively an analysis of the explanation of rational action; it is an analysis of the conceptual structure of intending, believing and being able to act.

It can now be shown, and this is of importance both for the theory of explanation and for the theory of determination of action, that the explanatory schema can incorporate and explicate most instances of *all* the following action explanations occurring in ordinary speech and in science.

Explanations in terms of
- intentions
- beliefs
- abilities
- opportunities
- wants

---

in a particular situation, the agent's belief is explicitly cited instead of her intention, or why an external reason is cited instead of the relevant intention. This problem area is now known as the pragmatics of explanations. For an exellent introduction see B.C. van Frassen [106].

17 Thus I accept von Wright's contention that an intentional explanation can be viewed as a logically conclusive argument. Given my analysis of intention above, this can be demonstrated formally. See my [63], p. 129. In chapter 5 of this book I put this conclusion into the context of the debate between Hempelians and idealists concerning the nature of explanation in the humanities. For an introduction to this discussion see S. Hook [39].

emotions
certain attitudes and character traits
sensations
perceptions
reasons
compulsion.

Let me briefly present the basic arguments for this general contention.

EXPLANATIONS IN TERMS OF INTENTIONS, BELIEFS, ABILITIES AND OPPORTUNITIES

It is easy to see that ordinary explanations in terms of intentions, beliefs, abilities and opportunities are only incomplete variants of the complete intentional explanans. Consider the following examples: (i) A went into the bookshop because she intended to buy a certain book. (ii) A went into the bookshop because she believed that the book was to be found there. (iii) A went into the bookshop because she believed that she had to do that in order to find the book. (iv) A went into the bookshop because she was capable of and unprevented from doing so.

Together these colloquial explanations form a complete intentional explanans. In a situation when only one is quoted it is easily seen that the others must be presupposed. When (i) is explicitly given (ii) - (iv) are implicitly taken to be the case. When (ii) is explicitly given (i), (iii) and (iv) are implicitly taken to be the case, etc. When, for instance we say that *A* went into the bookshop because she believed that a book was to be found there, we must certainly presuppose that she intended to get hold of this book and that she thought it was necessary to actually enter the shop in order to get hold of it.

EXPLANATIONS IN TERMS OF WANTS

My previous analysis of wants in terms of dispositions to form intentions makes it quite obvious how the locution "*A* did *F* because she wanted to bring about *P*" should be analysed. The analysis must proceed in two steps. In the first step a want explains the occurrence of an intention, given that the antecedents of the want-conditional are fulfilled. In the second step the intention that has been formed explains the action in the way analysed above.

EXPLANATIONS IN TERMS OF PERCEPTIONS AND REASONS: RATIONAL EXPLANATION

Actions are frequently explained, not by reference to any mental property of the agent but by reference to a fact external to the agent. Consider the following example: "*A* jumped on to the pavement because there was a car coming towards her." How is the relation between such an external fact and the human action to be analysed?

It is obvious that a fact such as a car coming towards an agent cannot be treated in the way that I have proposed for intentions and wants. There is no conditional element in it; it cannot be treated as a disposition. It is then very tempting to assume that the relation between an external fact and an action in the case where the former explains the latter is in a straightforward way causal.

Many contemporary philosophers have, however, firmly denied that this can be the case. A *reason* they claim, does not explain in the way that a cause does. In an analysis of an explanation in terms of reasons (or a rational explanation) we must take account of the peculiar relation that holds between the notions of reason and intention. There is, they say, no analogous relation between cause and intention.[18]

In the treatment which I shall propose, I shall attempt to explicate the nature of the relation between reasons, intentions and actions. I shall say that a reason is an explanatory factor which is, in contradistinction to most causes, a part of an intentional explanans or, in certain cases, a want explanans. More specifically, the reason, or strictly speaking the agent's awareness of the reason, constitutes the fulfilment of one of the conjuncts of the antecedent of an intention- or want-conditional.

Let me first analyse the concrete case. How could it be that the car's coming towards *A* "made" *A* jump on to the pavement? One trivial but necessary condition was that *A* became aware of the fact (or at least believed it to be a fact) that the car was approaching her. Another necessary condition was that *A* judged the situation to be what it was and judged that there were consequences it was likely to have — in short that *A* believed the situation to be dangerous. If *A* did not believe that the car was dangerous to her and that it would probably kill her, her observation of the car would obviously have nothing to do with the explanation of her movements.[19]

Still, this is not enough. There would be no action on the part of *A* in this situation unless there was something that *A* *intended* to achieve. *A* would not have moved unless she had had the intention (at least) to survive. If *A* had been in a mood where she did not care about her life or health, her observation of the dangerous car would not have incited her to react in the way she did. (I am of course here presupposing that we are not dealing with a pure reflex movement.) Since *A*, however, intended to preserve the state of being alive, she consequently set herself to jump out of the way. To jump away was, as she correctly judged it, a necessary and sufficient condition for staying alive, given the circumstances surrounding her. Consider the case schematically:

18 For an early discussion of rational explanation as contrasted with causal explanation, see W.H.Dray [25].

19 In a recent paper [116] G.H. von Wright deepens the discussion about explanations in terms of reasons. His main concern is how we differentiate between *existing* reasons for an action and *efficient* reasons for an action, the latter being the ones in terms of which we can explain the actions.

(i) There is a car coming towards *A* (the reason).
(ii) *A* becomes aware of (i) and judges it to be what it is.
(iii) *A* intends to avoid being killed.
(iv) *A* believes that she is going to get killed in the circumstances surrounding her, unless she jumps on to the pavement.
(v) *A* is capable of and not prevented from jumping on to the pavement.

(vi) *A* jumps on to the pavement.

Given this analysis we can see that the components nicely fit the pattern of the intentional explanation. The reason is identical with that circumstance which, when perceived by the agent, is considered by her to make a certain course of action necessary in order to achieve or preserve an intended end. The reason, then, is to be found in the second component of the intentional explanans. The fact that the agent perceives or otherwise becomes aware of the reason constitutes the whole second component of the explanans.

According to this analysis, then, a rational explanation is another elliptic way of explaining an action. It is now easily seen that the explanation "John jumped on to the pavement because he *saw* a car approaching him", i.e. an explanation in terms of a perception, follows exactly the same pattern.

An important special case of rational explanation is the one where the reason has the form of a *normative principle.* Consider, for instance, "*A* thanked the hostess for the dinner because that is the thing to do in Swedish society."

Which are the necessary links between the code of etiquette saying "When one has been invited for dinner, after the meal one should thank the hostess for it," and the fact that *A* at a particular time thanks a particular hostess for the dinner? How does the normative principle "make" *A* act in accordance with it? Again, a necessary condition is that *A* is aware of the principle and believes it to be what it is. A further condition required is that *A* considers the principle to be applicable to the kind of situation in which she is placed. This means that *A* must believe that, under the circumstances, she cannot conform to the principle unless she thanks the hostess for the dinner.

But again, there will be no action on the part of *A* unless she has the intention to conform to the principle in question. The situation may be schematised:

(i) There is a normative principle of the kind: "when one has been invited for dinner, after the meal one should thank the hostess for it."
(ii) *A* believes that she cannot conform to this principle, given the circumstances, unless she thanks the hostess for the dinner
(iv) *A* intends to conform to this principle
(v) *A* is capable of and not prevented from thanking the hostess for the dinner.

(vi) *A* thanks the hostess for the dinner.

A SPECIAL CASE OF RATIONAL EXPLANATION: COMPULSION

Consider now the explanation "The soldier shot at the enemy because he was forced to do so". The fact that an agent is compelled or forced to perform a certain action is sometimes cited as the explanation of the action. This case has, I shall claim, its place in the context of rational explanation.

An agent is said to have been compelled to perform a certain action *F*, when there is an external "force" (sometimes also internal) such that *A* "could not have avoided" doing *F*. The primary use of these locutions is in the field of interaction between agents. One agent *B* compels another agent *A* to perform *F*, e.g. by threatening, blackmailing or simply ordering her. In a derivative use of the word, however, compulsion may also be a relation between inanimate objects or an inanimate object and an agent. For instance, "the river forced *A* to go a roundabout way."

On the surface of it this kind of explanation seems to be very far from rational or even intentional explanation. A person who is forced is said to "have no choice". I shall, however, challenge this *prima facie* impression. A compulsion or a force is, I claim, an external reason for an agent to perform a certain action. This may be brought out in the following analysis. *A* is a private in the Swedish army during wartime. *A* is ordered by his officer to shoot at an enemy platoon approaching his position. *A* knows that if he does not shoot, he will be court-martialled and in the end be sentenced to death. *A*, however, intends to survive. Hence, he will set himself to shoot at the enemy.[20]

This case is in all essentials on a par with the case of the approaching car. The only slight difference lies in the surface locutions. In the case of the approaching car the reason is spelt out in detail. In the force-case the locution can sometimes be quite elliptic: "He acted because he was forced to do so."

What I have analysed here is a case of forcing somebody to perform an intentional action. The terms "force" and "compulsion" are sometimes used for the cases where the consequent is not an action but an event caused by an action of physical force. A salient example is the case where a father drags his son home from a disco. Here the son is forced but not in the way that he is made to perform an intentional action.

It could then, however, be asked: why are some reasons considered to be compelling reasons and some not? How is that distinction to be clarified? *A reason is compelling,* I shall suggest, *when it presupposes, and threatens the realisation of, an intention which is such that the agent is not prepared to relinquish it.* Very few agents are, for instance,

20 In the very significant work *Coercion* [108] Alan Wertheimer emphasises the central position of threat in coercion. I agree with Wertheimer that threat is central in many instances of coercion. My analysis of threat, however, differs significantly from Wertheimer's. According to Wertheimer a threat should be characterised in the following way: "*A* threatens *B* by proposing to make *B* worse off relative to some baseline" (p. 204). The differences between our positions are the following:

(i) Wertheimer limits himself to the case of coercion where there is a human agent who *intentionally* threatens another agent. I wish to talk about compulsion or coercion much more generally.

(ii) I emphasise the perspective of the agent coerced. In order for *A* to be compelled or coerced *A* must consider that a very basic intention of hers is threatened with not being realised.

(iii) I suggest a further element in the genuine case of coercion. The action taken by the compelled agent is considered by her to be necessary because she is not prepared to relinquish the intention whose realisation is threatened.

prepared to relinquish their intention to survive. Hence every reason which threatens the realisation of such an intention will be considered as a compelling reason for the agent. (For further illustrations of compulsion, see chapters 6 and 12. My reconstruction of force or compulsion also has important implications for my analysis of accountability in criminal law. See chapter 14.)

EXPLANATIONS IN TERMS OF CHARACTER TRAITS AND ATTITUDES

Consider the example: "*A* joined the army because he is patriotic". The fact that *A* is patriotic, i.e. the fact that he has a certain kind of ideological attitude, is here cited as the explanans of an action of his, viz. his joining the army. How is this explanation to be analysed? To what kind of entity does one refer when one refers to an attitude?

In general, to ascribe an attitude or a trait to a person is to ascribe to her a capacity, a liability or tendency, in short a disposition to behave or act a certain way. Quite frequently and typically the disposition (or set of dispositions) which is entailed by a certain attitude- or trait-concept is a want or an intention. Examples of such concepts are patriotism, vanity, egoism, kindliness, ambition and philanthropy. To be vain is tantamount to wanting or intending to show off and be the object of admiration; to be egoistic is tantamount to wanting or intending to satisfy one's own interests before the interests of others; to be kind is tantamount to wanting or intending to help and support one's fellow human beings, etc.

An explanation in terms of such properties will then receive the following standard analysis:

(i) *A* is patriotic, which means among other things,
A intends to support his native country.
(ii) *A* believes that he cannot support his native country, given the circumstances, unless he joins the army.
———
(iii) *A* joins the army.

Consequently, an explanation in terms of such an attitude or character trait will become an incomplete variant of an intentional explanation in the same way as ordinary explanations referring explicitly only to intentions are incomplete variants.

I wish to emphasise here, though, that this analysis cannot be universally applied to the field of attitudes and character traits. Some of these entail capacities or dispositions of other kinds. Consider, for instance, irascibility and stubbornness. These are properties of an individual which need not entail intentions or wants.

EXPLANATIONS IN TERMS OF EMOTIONS AND SENSATIONS

Analogies and disanalogies between emotions and sensations. Emotions, in particular, but also sensations, may be cited as explanatory factors in respect of actions. Since these two kinds of entities are often confused, it might be illuminating to scrutinise their nature in a comparative way. In performing this comparison I shall ally myself with an analysis of emotions (and other mental phenomena) developed in particular by A. Kenny [43] and later by I. Pörn [80] and O.H. Green [34].

From a common-sense point of view emotions and sensations seem to have at least two traits in common:

(i) They both in one way or the other involve feelings.
(ii) They both typically have bodily expressions.

(i) According to one common way of speaking, emotions and sensations are simply two kinds of feelings. To have an emotion or to have a sensation is merely to have a certain kind of feeling, where the emotive feeling is perhaps distinguished from the sensational one by embracing the totality of the person, by not being located in any particular part of the body.

This mode of speech is however not very clarifying. In my opinion only sensations can be candidates for being equated with feelings. That this is so can be illustrated by the following consideration. It makes sense to say that *A* has a certain kind of feeling at every moment in which it is true to say of her that she is in pain (i.e. has a particular kind of sensation). It does not, however, make sense to say that *A* has a certain kind of feeling at every moment at which it is true to say of her that she is in love (i.e. has a particular kind of emotion). It may be true to say that *A* is in love with *B,* even while *A* is asleep and hence cannot have any feelings whatsoever. This is not said in order to deny that emotions also typically *involve* feelings, for instance that one sometimes has a particular kind of feeling when one is in love. Perhaps it is also true to say that one cannot be said to be in love if one never has this feeling.

(ii) Emotions and sensations are also alike in the respect that they typically have physical expressions, some of which are behavioural. Crying is a behavioural expression of being in pain; turning pale is a non-behavioural expression of the same sensation. Trembling is a typical behavioural expression of the emotion anger, whereas blushing is a standard non-behavioural expression of the emotion embarrassment.

Emotions and sensations are, however, obviously unlike in several respects. In short, emotion-concepts are much more complicated than sensation-concepts.

(i) As I have already indicated, emotions are not, like sensations, restricted to any particular part of the human body. One feels pain in one's leg, but one does not feel love in one's leg or any other particular part of the body. This is not to deny that some of those feelings which are connected with certain emotions can have a fairly clear location. When one is e.g. nervous about something, one may have a sinking feeling in one's stomach. This does not, however, give us reason to say that one feels nervousness in one's stomach.

(ii) Most emotions are conceptually connected to certain situational facts which are often called the *reasons* for the emotions. Some emotions can occur, and this for conceptual reasons, in certain situational surroundings only. A person can e.g. be grateful only in the context where she at least believes that some other person has contributed to her welfare or the welfare of someone near to her.

(iii) This fact in turn explains why emotions are normally directed to objects outside the agent. Consider:

(1) *A* fears the approaching car.
(2) *A* envies the famous film star.

Sensations like "being in pain" and "feeling warm" are obviously not directed in this way, nor do they have reasons. The connection between objects of emotions and reasons for emotions can be indicated by the following observations: the approaching car, which is the object of the emotion fear in our example, is also part of that environment which consitutes the reason for *A*'s fear. The film star, the object of *A*'s envy, is that person whose various successes constitute the reason for *A*'s envy.

(iv) For my purposes it is important to note that many emotions to be found among those which occur in explanatory contexts are conceptually tied to certain intentions (or wants) as well.

It is logically impossible to be afraid of something *x*, if one does not believe that *x* is a threat to one or more of one's most basic intentions or wants. When *A* is, for instance, afraid of an approaching car, that is so because *A* has a standing intention to protect her life and health. If *A* did not have this or a similar intention, it would be inadequate to ascribe the emotion of fear to her.

As a summary of these observations about emotions I shall propose the following specific analyses of the emotions of *fear* and *envy.*

(a) *A* is afraid only if
  (i) *A* is disposed to have a particular kind of feeling.
  (ii) *A* is disposed to have certain bodily expressions.
  (iii) *A* believes that there is an event occurring (the reason) which is such that it threatens one or more of her most basic intentions or wants.

(b) *A* is envious only if
  (i) *A* is disposed to have a particular kind of feeling.
  (ii) *A* is disposed to have certain bodily expressions.
  (iii) *A* believes that there is another agent *B* (or a class of agents *B1*,....,*Bn*) who has (have) gained great success, privileges, money or other desirable things (the reason), a fact which threatens a basic intention or want of *A*'s. The intention or want is in this case rather specific, perhaps: *A* intends (or wants) to create (or protect) a state of affairs such that no one (at least no one in *A*'s immediate environment) appears as a more prominent person than *A* herself.

EXPLANATIONS IN TERMS OF EMOTIONS

The proposed analyses of our paradigm examples of emotions seem to give sufficient clues for the clarification of explanations in terms of them. (I shall here confine myself to discussing the case where the analysis can be made in terms of intentions.) Consider: "*A* jumped on to the pavement because she was afraid of a car that was approaching her," and the following analysis:

(1) There is a car approaching *A*.
(2) *A* is aware of this fact.
(3) *A* is afraid of the car (which partly means:)
(3') *A* has an intention which is, *A* believes, threatened by the fact that the car is approaching her, i.e. *A* intends to survive.
(4) *A* believes that, given the circumstances, she is not going to survive unless she jumps aside.

---

(5) *A* jumps aside.

What, then, is the specific function of citing an emotion-word in an explanation? Our proposed analysis seems to be almost identical with the analysis of rational explanation. On the one hand, these analyses become similar for the simple reason that an explanation in terms of an emotion of the type fear is, in analogy with rational explanation, an incomplete variant of intentional explanation. On the other hand, by citing an emotion one makes explicit certain aspects of the explanans other than those made explicit by citing a reason. Consider the following two cases:

(i) If the ordinary-language explanans is "because *A* was afraid of the approaching car," one makes explicit not only that there is an intention on the part of *A* which *A* believes to be threatened but also that the circumstantial fact which threatens it is the fact that a car is approaching her.

(ii) On the other hand, if the ordinary-language explanans is merely "because *A* was afraid," the circumstantial fact is not specified; it is merely indicated that there is a fact which is such that it threatens a basic intention of *A*'s.

We can now suggest an analogous treatment of, for example:

A tried to entrap *B* in a scandal because *A* envied *B*'s successes.

By stating "because *A* envied *B*'s successes," one makes explicit that the reason for *A* to act in the way she did was the fact that *B* had had great successes in the past and that this fact threatened *A*'s intention to appear as a prominent person in the circles where she lived.

In general, to explain an action by citing an emotion is in these typical cases tantamount to explaining the action by (more or less) specifying the reason and the operative intention behind the action. It is also interesting to observe that what I have called the reason for the emotion is identical with the reason for that action which is explained in terms of the emotion. The reason why *A* is afraid of the car is also the reason why *A* avoids the car.

At this stage it is natural to ask: does one not say something more about the agent when one ascribes fear to her than when one merely points out her intentions and situational surroundings? I agree that an emotive word carries a greater load. Among other things it indicates, as I have said, that the agent is liable to have particular kinds of feeling. I do not, however, think that the study of such features contributes to clarifying the explanatory force of emotions.

EXPLANATIONS IN TERMS OF SENSATIONS

Sensations are not frequently cited as explanatory factors in respect of actions. They constitute rather the typical explanations of such behavioural expressions as are non-actions, e.g. groaning and crying. It seems, however, quite possible that sensations may also explain certain intentional actions. It looks, for example, as though people may *decide* to perform certain actions because they are in pain. A natural suggestion in order to handle such a case is to say that a more complete analysis of the sensation "being in pain" should also contain certain dispositions to form intentions. I think, however, that this is an idea inadequate for sensations.

When sensations are cited as explanantia of intentional actions, they explain, I shall suggest, by being *reasons* for decisions or *reasons* for actions. In other words, they explain against a background of already existing wants or intentions on the part of the agent. How this can be so is perhaps clearly seen in the following example: "*A* went to the hospital because she was in great pain," the analysis of which would be:

(1) *A* is in pain.
(2) *A* intends to live a pleasant life.
(3) *A* believes that, given the circumstances, she cannot live a pleasant life, (i.e., rid herself of the pain) unless she goes to the hospital.

---

(4) *A* goes to the hospital.

It is, of course, absurd to maintain that going to the hospital is an expression of being in pain. Still, a sensation like being in pain is a possible explanans of an action such as going to the hospital.

It is important to note that this model also is a possible but very untypical model for explanations in terms of the previously mentioned mental properties like character traits, attitudes and emotions. If for example *A* went to the psychiatrist because she is very sensitive or nervous, then her sensitivity or nervousness would explain her action by being a reason for taking action. It would not explain according to the standard models for explanations in terms of traits, attitudes or emotions.

AN UNTYPICAL EXPLANATION IN TERMS OF PERCEPTIONS

I have previously treated explanations in terms of perceptions as a subspecies of rational explanation. That is indeed the typical case. "*A* called for the ambulance because she had seen a car-crash" is normally equivalent to "*A* called for the ambulance because there had been a car-crash." The particular fact that the agent has become *aware* of the reason, i.e. has observed it, which is indicated by the former but not by the latter example, must be a part of the analysis of the latter explanation as well.

But perceptions may also explain in a slightly more subtle way. It need not be the fact perceived which functions as the reason; it may be *the fact that the agent perceives a certain fact* that is the reason. Consider, for example, "*A* fled from the temple because she had seen the god." If we assume that it is an essential part of *A*'s religion that she is not allowed to see the god, the reason for *A*'s action need not be that there was a god to be seen in the temple but the very fact that she, *A*, had caught sight of the god.

In this case, the explanation in terms of a perception is an even more elliptic variant of an intentional deterministic explanation. Here the only explicit part is *C* of the conjunct: *A* is convinced that *C*. The complicating fact, though, is that *C* itself is of the form: *A* is convinced that *C*.

SUMMARY

We have found that explanations of actions in terms of character traits, attitudes, emotions, sensations and perceptions may all be incomplete variants of intentional explanation. They are, however, variants incomplete in slightly different ways: in the case of

- (i) traits or attitudes, the explicit part is the intention itself;
- (ii) emotions, the explicit part is the intention and the reason (both may be only vaguely specified);
- (iii) sensations, the explicit part is the reason;
- (iv) perceptions, the explicit part is the agent's awareness of the reason.

In non-standard cases we noted that traits, attitudes, emotions and perceptions may also explain by being reasons for actions. Some of the mentioned mental factors, in particular traits and sensations, may be cited to explain unintentional behaviour as well. They then explain according to the paradigm of ordinary dispositional explanation.

ON THE NOTION OF AN INTENTIONAL HIERARCHY

Up to this point I have been assuming that there is one and only one intentional explanans to be found for a particular action. A simple reflection, however, indicates that there may sometimes be a choice between several explanantia all of which are correct from a formal point view. Here, we must distinguish between two important cases. The first is the case of *overdetermination*.

An agent may perform F both for the reason that she intends to bring about P and for the reason that she intends to bring about some other state of affairs, say R, for whose achievement she also and independently considers the doing of F necessary. The action F is then said to be overdetermined. One of the two intentions alone would be sufficient for A's performance of F.

The second case, which reveals much more of the features particular to intentions, is a case where two competing intentions are not independent of each other. Consider the following two explanations of one and the same action:

(1) *A* went into the bookshop because she intended to buy a book.
(2) *A* went into the bookshop because she intended to give her husband a birthday present.

These two explanations may both be true without there being a case of overdetermination. The reason for this is that, although the two intentions are distinct, they need not be and would not naturally be logically unrelated. The connection between the two explanations can in a preliminary way be brought out in the following complex proposition:

(3) *A* went into the bookshop because she intended to buy a book, *and* she intended to buy a book because she intended to give her husband a birthday present.

Hence what I suggest here is that the intention which is involved in (2) does not merely explain the action in question but also explains the intention which is involved in (1). To be more precise, we could say that the intention in (2) explains the action through the explanation of the intention in (1).

This indicates that we can construct a hierarchy of intentions related to each other and finally related to an action in the following way:

(1) *A* intends to bring about P (Level I)
(2) *A* believes that she cannot bring about P unless she brings about R.
____
(3) *A* intends to bring about R. (Level II)
(4) *A* believes that she cannot bring about R unless she does F
____
(5) *A* does F. (Level III)

It might now be asked: why is there such a hierarchy of intentions and how is it more fully to be understood?

In general, when an intention is very abstract or far-reaching in time or space, there is no simple action or set of actions that is necessary and sufficient for the achievement of the intended end. The "long" way that has to be covered to obtain the end may have to be divided into a number of substages, the achievement of which is necessary and jointly sufficient for achieving the main end. If one intends to achieve the main end, one must also intend to achieve the subends and the subends of those subends, and so on.

Consider a man *A* who intends to travel from Tokyo to Kvickjokk in Sweden. Assume that in order to do this *A* has to do the following: get to Stockholm by plane, get from Stockholm to Kiruna by train, and finally get to Kvickjokk from Kiruna by bus. Choose the first substage, i.e. travelling from Tokyo to Stockholm by air, and find the necessary and jointly sufficient conditions for achieving that are: to book an air-ticket in Tokyo, to get to Tokyo Airport and finally enter the relevant plane. A complex explanation of the fact that *A* books an air-ticket in Tokyo could, then, run thus:

(1) *A* intends to travel to Kvickjokk (Level I)
(2) *A* believes that he cannot get to Kvickjokk unless he travels by air to Stockholm
____
(3) *A* intends to travel by air to Stockholm (Level II)
(4) *A* believes that he cannot get to Stockholm unless he books a ticket in Tokyo
____
(5) *A* books a ticket in Tokyo (Level III)

What, then, are the limits for this hierarchy? What is the nature of the highest- and the lowest-level intention? Where does the action enter into the picture?

Trivially put, the highest-level intention of a particular hierarchy is such an intention for which there is no higher-level intention from which it can be derived in the manner demonstrated above. It is that intention which, as we would say, ultimately determines the agent's choice of action. Such an intention is normally of an extremely abstract or far-reaching kind. Typical examples are: the intention to survive, the intention to live a good life. For such intentions there would rarely be explanations in terms of higher-level intentions.

At the other end, the intentional hierarchy is limited by the basic actions. There cannot be an intention of a lower level than an intention to perform a basic action. The reason for this is obvious. The only necessary means for performing a basic action is the basic action itself. Hence, an intention derived from an intention to perform a basic action cannot be distinct from the latter intention. To illustrate this fact, consider the following derivation from an assumed highest-level intention to a basic action (for this example I have chosen necessary means, which also happen to be sufficient):

(1) *A* intends to save the republic (Level I)
(2) *A* believes that he is not going to save the republic unless he kills the president
____
(3) *A* intends to kill the president (Level II)
(4) *A* believes that he is not going to kill the president unless he shoots him with a gun
____
(5) *A* intends to shoot the president with his gun (Level III)
(6) *A* believes that he is not going to shoot the president unless he pulls the trigger of his gun
____
(7) *A* intends to pull the trigger of his gun (Level IV)
(8) *A* believes that he is not going to pull the trigger of his gun unless he moves his hand
____

(9) *A* intends to move his hand (Level V)
(10) *A* believes that he is not going to move his hand unless he moves his hand
(11) *A* is capable of and not prevented from moving his hand
(12) *A* moves his hand. (Level VI)

By moving his hand *A* pulls the trigger, by pulling the trigger *A* shoots the president, by shooting the president *A* kills him, and, finally, by killing the president *A* saves the republic.

Since it is a truth that for every non-basic action there is a basic action, it is easily seen that for every case of intentional action there is a complete derivation down to the lowest-level intention to be constructed. Strictly speaking, an agent does not "proceed" to action until she has gone through the whole sequence down to the basic action. (By this I do not, of course, mean that she has to make a conscious deliberation in accordance with this sequence of inference.) The following explanation could then be regarded as an *elliptic* formulation of the structure (1)-(12):

(13) *A* intends to save the republic
(14) *A* believes that he is not going to save the republic unless he kills the president
(15) *A* kills the president.

Objections to the Proposed Analysis of the Concepts of Intention and Intentional Explanation

There are certain important prima facie arguments to be directed against the analysis of the concepts of intention and intentional explanation as presented above. The most obvious and important ones stress the fact that there are, or at least seem to be, cases where an agent has an intention to bring about a state of affairs *P*, considers an action *F* necessary for the realisation of *P*, can do *F*, but nevertheless abstains from doing *F*.

(i) *A* may abstain from doing *F* because, although she can do *F*, she finds that she is not capable of performing one or more of those other actions which she considers necessary for achieving *P*. (The argument from incapacity.)

(ii) *A* may abstain from doing *F* because her intention is withdrawn or otherwise dropped before she has been able to do *F*. (The argument from time.)

(iii) *A* may abstain from doing *F* because she has some other "stronger" intention requiring that she shall not do *F*. (The argument from conflicting intentions.)

(i) The argument from incapacity

Consider the following case: *A* intends to cross a river. *A* knows that she will not be able to cross the river unless she walks down to the river bank. Hence, unless *A* is prevented or incapable she should set herself to walk down to the river bank.

Nevertheless, she may not do so for the reason that she believes there is no other action which is performable by her and which in conjunction with walking down to the river bank forms a sufficient condition for getting to the other side of the river. It may be that there is no bridge or boat available and that it happens to be impossible to cross the river by swimming. When this is so, it is of no use for *A* to walk down to the river, as she would still not achieve her goal.

The answer to this objection is rather simple: as soon as *A* realises or believes that she is not capable of achieving the intended end, then she does not any longer, *per definition,* intend to achieve it. According to our analysis, one can intend to bring about *P* only if one believes that one can bring about *P,* i.e. that one can bring about all those subends and perform all those actions which one considers to be necessary for the achievement of *P.*

(ii) The argument from time

Selfevidently, there is in an intention a reference to a state of affairs which is assumed to occur at a *future* time. In order to more clearly illustrate this fact we may expand "*A* intends to bring about *P*" and "*A* believes that her doing *F* is necessary for the bringing about of *P*" into:

(1) *A* (now) at *t* intends to bring about *P* at *t2*.
(2) *A* (now) at *t* believes that she cannot bring about *P* at *t2* unless she does *F* at *t1*.

Quite frequently the state of affairs that *A* intends to bring about is placed (or rather assumed to be placed) in the distant future. *A* may now intend to become president in five years and *A* may now believe that she will not become president in five years unless she makes a certain number of political speeches during the fourth year of this period. Facing a case like this, it is easy to imagine the following: during the third year, *A* suddenly drops her intention to become president. She may perhaps have realised that no one was willing to support her in her aspiration. Hence she is of course not going to make any speeches the year after.

Nevertheless it is true to say of *A* that during a period of three years she had been intending to become president, that she during the same period believed that she would not become president unless she made some political speeches during the fourth year, that she was quite capable of and not prevented from making these speeches and that she still in the end did not do that.

This is a possible case but it cannot, as far as I can see, be counted as a serious argument against our analysis. What the argument in fact illustrates is the case where a law ceases to be a valid law and as a consequence cannot govern the course of events. Consider the following parallel: assume that there is a bomb with such a mechanism that if the bomb is primed at *t* it will explode at *t10*. Hence, there is a law holding true of this bomb saying: if the bomb is primed at *t,* it will explode at *t10*.

Now, it is of course possible for someone or something to disturb or even destroy the mechanism of this bomb, an action which will have as a consequence that the bomb will not explode at *t10*. The law that has hitherto governed the bomb has then ceased to be valid.

The possibility of the fact that a law may cease to be a law is, however, hardly an argument for saying that the standard analysis of laws is incorrect. Moreover, in an explanatory argument using a law it is certainly taken for granted that the law existed as a law at least to the time of the occurrence of the explanandum. Analogously, in those explanatory arguments where an intention (which is also, as I argue in chapter 5, partly to be viewed as a kind of law) is among the premises, it must be presupposed that the intention existed at least up to the time of the action to be explained.

(iii) The argument from conflicting intentions

A's intention to bring about *P*, it may be argued, can fail to result in an action *F* which is considered by *A* to be necessary for *P*, for yet another kind of reason. Consider the following case:

A intends to become president. *A* believes that she is not going to become president unless she defames her political antagonists in her speeches. At the same time, however, *A* has the intention to live a moral life. She believes that she is not going to live a moral life unless she abstains from defaming her fellow human beings. Assume, now, that *A*'s intention to live a moral life is an intention which is "stronger" than her intention to become president. As a consequence of this, *A* abstains from publicly defaming her opponents.

This kind of case does not, however, seem to be consistent with our analysis of intentions. It still seems to be true that *A* intends to become president, considers defaming her antagonists to be a necessary means for becoming president, and is capable of and not prevented from doing so. Hence our theory requires that she sets herself to act according to this intention.

My suggestion to resolve this problem is the following. I claim that there cannot, strictly speaking, be such things as conflicting intentions where the agent is aware of the conflict. The reason is that if the agent were to be aware of such a conflict, then she would be aware of the fact that she couldn't perform both actions. But having an intention to perform *F* entails, per definition, believing that one is capable of performing *F*. Hence, the agent cannot have two intentions which are in conflict with each other. When one talks, as in the above examples, about conflicting intentions, this is in actual fact a misuse of language. *A* may, one might argue, have a *want* which is in conflict with an intention of hers, but she cannot have two intentions standing in conflict with each other.

CHAPTER FIVE

ON THE LOGICAL FORM OF ACTION-EXPLANATIONS

PRELIMINARIES ON INTENTIONAL ACTION EXPLANATION

Let us recall the partial characterisation of intention proposed above:

A intends to bring about *P*, only if

A is in a state which is such that
(i) *A* believes that she can bring about *P*, and
(ii) For all actions *X*, if *A* believes that she cannot bring about *P* unless she does *X*, and if *A* can do *X*, then *A* will do *X*.

Combine this definition with the schema for intentional explanation:

(i) *A* intends to bring about *P*
(ii) *A* believes that she is in situation *S*
(iii) *A* believes that in *S* she will not bring about *P* unless she does *F*
(iv) *A* can do *F* in the situation in which she is.

Replace (i) in the schema by its analysis above. It then follows as a logically valid conclusion that:

(v) *A* does *F*.

This result has important consequences for the resolution of a debate which was conducted in analytic philosophy mainly during the 60s and 70s concerning the logical character of action-explanations. A starting poing for this discussion was the analysis proposed by the logical positivists, in particular Carl Hempel.[21] In a number of important works Hempel argued that a scientific explanation must consist of the following elements: a universal law premise, an initial condition premise, and the explanandum (or, strictly speaking, a sentence expressing the explanandum). Formally:

$$\frac{\begin{array}{l} L1 \ldots Ln \\ C1 \ldots Cn \end{array}}{E}$$

21 The best introduction to Hempel's philosophy of explanation is his *Aspect of Scientific Explanation and Other Essays in the Philosophy of Science* [36].

The sentence expressing the explanandum should be deducible from the two premises. The model thus described is the famous deductive-nomological model of explanation. This model has been followed by several slightly different versions (formulated by Hempel himself), some of which contain probabilistic laws instead of universal deterministic laws. A common feature of the various explanantia, however, is that they contain some law-like premises.

A number of writers with a humanistic inclination contested the applicability of this model, in any of its versions, to the explanation of human actions. They said that there is something special about explanations of human actions because there is a peculiar relation between the typical explanantia of human actions and the human actions themselves. A typical explanans of a human action, they said, is an intention or a reason. One explains why a person does something by referring to her intentions or reasons. This kind of explanation, the argument goes, is not an ordinary causal explanation of the kind that we meet in the natural sciences. An intention or a reason cannot be a cause of the action that it explains.

The latter crucial contention is in its turn backed up by different arguments from different authors. The historian William Dray [25] focused on reasons and what he called rational explanations. He claimed that the most significant feature of a rational action-explanation is that it shows the *appropriateness* of the action to be explained. This distinguishes rational explanations from causal explanations. When one explains the movement of a billiard ball by citing its cause, one does not in any sense show the appropriateness of the movement of the ball. Other authors, for example Abe Melden [56] and G.H. von Wright [114], have advanced the thesis that the relation between the intention and the intended action is in some sense logical. And a logical connection between two entities excludes an empirical causal connection between the same entities.[22]

Moreover, the humanistic theorists claim, there is in the explanation of action hardly ever any reference, explicit or implicit, to a general law. When one explains why Peter does something, one normally does not refer to any information about how people who are similar to Peter behave in a similar situation. Instead one wants to know further things about Peter, what he intended in the situation in question, what he knew or believed about possible means to fulfil his intention. In short, according to the humanistic lines of reasoning, intentional or rational explanations are non-causal and non-Hempelian. These explanations are peculiar and distinct from ordinary scientific explanation.

My purpose in the following is to assess this critique, in the light of my own analysis of the notion of intention. I wish to account for the insights of the humanists (and I shall then in particular focus on the idea of a logical connection) but I also wish to show the basic similarities between the schema for intentional explanation and the general schema of deductive-nomological explanation.

As a preparation I shall make the following observations.

(i) The schema for deductive-nomological explanation is a logical argument. The explanandum follows logically from the explanans. So the fact that a sentence describing

---

22 W.H. Dray presents his theory of rational action-explanation in [25], pp. 118-155.

an action to be explained follows logically from a conjunction of explanans-sentences need not indicate anything peculiar about action-explanation.

(ii) Observe also that logical conclusiveness can be obtained in an explanatory schema without the help of a universal premise. This holds for instance for so-called dispositional explanations.

The notion of a *disposition* will in fact be a key notion in my account. My whole analysis relies on the hypothesis (advanced and developed in chapters 2 and 4) that intentions can be viewed as a species of dispositions.

Consider first a simple kind of disposition, viz. brittleness, and an explanation in terms of it. We wish, for instance, to explain that a window breaks by referring to its dispositional property brittleness and, in addition, to the fact that something hard hit the window.

An explanation of the fact that the window breaks can then be put in the following way: The window breaks because the window was hit by a hard object *and* because of that property of the window, viz. the brittleness, which is such that if the window is hit by a hard object, then it breaks. In this formulation it is evident that the explanandum follows logically from the explanans. The logical structure is the following:

$$\frac{\begin{array}{l} Fa \rightarrow Ga \\ Fa \end{array}}{Ga}$$

We observe that this is an explanation which is deductive in form, hence fulfilling one of Hempel's requirements regarding an explanation. Still, it is not deductive in virtue of the existence of a universal law. It is deductive in virtue of a dispositional hypothetical.[23]

This, then, is in principle the simple logic that can be applied to the intentional explanatory schema put forward above. I have in fact proposed a dispositional analysis of intentions, and it is easy to show that the schema of intentional explanation is deductive in the way the ordinary dispositional paradigm is.

When it comes to details the situation is slightly more complicated with intentions. In the simple dispositional paradigm the consequent of the hypothetical is identical with the conclusion of the argument. In the intentional case there is normally no direct reference to the action to be explained. One intends to bring about a goal, a state of affairs *P*, and finally one performs an action *F* distinct from *P* but which is believed to be conducive to *P*.

23 There are various versions of the logical connection thesis concerning intentional action-explanation. A.I. Melden [56] denies the possibility of a causal relation between intentions and actions for the reason that an intention necessarily contains a reference to the action which it explains. G.H. von Wright [114] refers instead to the fact that the typical action-explanation has the form of a "practical syllogism". I owe this term to von Wright [114], p. 96.

An intention, as my analysis has suggested, is a disposition not just to perform a single action but to perform a number of actions, namely all those actions which the agent believes to be necessary for the achievement of the intended state. Hence in the intentional analysis we need some universal quantification, which is absent in the simple dispositional paradigm. But observe: this does not transform the expression into a universal law. The expression still refers to a particular agent *A* and no other agent; it is only that *A* is disposed to perform more than one action when she has an intention. She is disposed to perform *all* actions which have a particular property.

From this reasoning the following conclusions can be drawn:

(i) The schema of intentional explanation has the form of a deductive logical argument.
(ii) It contains no universal law, however. It is deductive in virtue of the existence of a dispositional hypothetical.
(iii) But since dispositional explanations occur outside the field of action-explanations, the intentional explanatory schema is in a sense reduced to a kind of explanation which is not peculiar to the humanities.

It could now, however, be asked what implication this solution has for the anti-causal idea. Does it still mean that intentions are not causes as the proponents of the logical connection argument insist?

The question requires a careful answer. I shall now first give only a partial answer, then return to the issue in replying to an objection to my theory. It should here be borne in mind that in the intentional explanation, as reconstructed by me, I do not use the whole content of what it means to have an intention. As I have mentioned earlier, to intend also means to believe that one can realise the intended end. This information is not necessary for the deduction of our explanandum. There may be other elements contained in the analysis of "*A* intends to bring about *P*". Likewise, there are other elements than the ones used in a dispositional explanation contained in the full analysis of, for instance, "this window is brittle".

This should be borne in mind because the answer that I give now concerns only that part of the analysis of dispositions and intentions which is contained in the explanatory arguments, viz. the hypotheticals. And the question can now be framed: Can a hypothetical fact be a cause?

How do we look upon things in the case of a Hempelian explanatory argument? What things are causes and what are not causes among the premises in this argument? If we deal with a Hempelian reconstruction of a causal explanation it is obvious that the cause or the causes are the factors which are put among the initial conditions. The universal law itself, the universal hypothetical, is not a cause, although it is of course one of the premises in the Hempelian reconstruction of a causal explanation.

Likewise in a dispositional explanation, in general, it is the second premise, or that which is signified by the second premise, which is considered to be the cause. In our brittleness case, it is the fact that the window is hit which is the cause. The hypothetical fact used in the dispositional explanation is not of the right category to be a cause. Hence the answer to our question whether the dispositional hypothetical is a cause is:

no. This means also that the *intentional* hypothetical is not of the right type to be a cause.

This conclusion, however, does not have anything to do with a logical-connection argument. It is a conclusion based on considerations about the ontological status of laws and hypothetical facts. (One ought to be able to distinguish between that which is a cause, and that which the cause is a cause in virtue of, viz. a law or a dispositional hypothetical. In *Explanation of Human Actions* [63] I in fact coined the term "semi-universal law" for the dispositional hypothetical fact.) But observe again, the hypothetical is not the whole analysis of the disposition. The hypothetical is true in virtue of a categorical fact. Nothing prevents this categorical fact from being a cause. I shall return to this issue below.

A DEFENCE OF THE THEORY

Consider now some basic questions that can be raised concerning my analysis: *First, by what arguments can I back the supposition that intentions are a kind of dispositions?*

The intuitive tests by which I can support the dispositional nature of intentions are identical with or similar to tests or arguments put forward by proponents of the logical connection thesis. Some of them, in particular von Wright [ ], ask: How do we confirm and how do we falsify hypotheses about intentions? Well, they say, we do this by looking at the agent's actions. If we know the beliefs of a particular agent and have a hypothesis about her intentions, then her subsequent actions or non-actions function as the test.

Assume that we put forward the hypothesis that M. Jacques Chirac has the intention to be renominated as a candidate for the presidency in France. We assume that Chirac knows all the measures that have to be taken in order for this situation to be realised. He must, for instance, see to it that a re-election committee is established. Assume also that we know that Chirac is both physically and mentally capable of taking these measures.

How do we test our hypothesis about the intention? Obviously, we would ask Chirac or we could ask his associates. But the answers to these questions would not be conclusive. The people involved may not know or they could lie, or perhaps abstain from answering.

The ultimate answer could only be given by the observation of Chirac's own actions. Does he perform the actions that he would consider necessary for his goal or not? If he does, we have confirmed (but of course not completely proved) the hypothesis. If he does not, then we have disproved the hypothesis. Because, if he knows how to act and is able to act accordingly but in fact does not act, then it cannot be true that he really intends to become the president again. He may of course still have an idle wish or he may hesitatingly want. But he cannot really *intend.* So the argument goes.

I find this argument a strong one. I think it is a striking analysis of our ordinary notion of intention. But of course it relies on my intuitions about the use of language. And I can obviously never convince someone who claims that her intuitions are different from these.

The second question is the following:

Is there not a striking difference between the paradigm dispositional explanation and the intentional explanation? The intentional explanation contains a clause about the agent's capabilities. The dispositional paradigm does not seem to have anything equivalent.

To this I will answer that the difference is just apparent. The dispositional paradigm as I have presented it is incomplete. Of course, the dispositional hypothetical does not hold true under all kinds of circumstances. In that hypothetical there is, as in fact there is in all universal laws, a *ceteris paribus* clause built in. A brittle window does not necessarily break when something hard hits it. There may be one or more preventing factors present in a particular situation — for instance the window may be strengthened by some support behind it, etc.

Hence I would say that in all detailed and careful formulations of dispositional explanations there must be a *ceteris paribus* clause present, both in the antecedent of the hypothetical and as a separate premise indicating that there is nothing preventing the initial condition from operating. If this is true, the intentional explanatory schema is not particular in this respect.

Thirdly, is there not a striking difference between intentions and ordinary dispositions in that the former come and go, normally existing only for a short period of time, whereas dispositions are lasting attributes of objects?

As far as I can see, there is no difference in principle here. It may be true that most of the celebrated examples of dispositions are lasting properties of objects. Brittleness certainly is. But of course they need not be. Another celebrated natural dispositional property, the property of being magnetic, can be temporary. An object may be magnetic for a short while and then lose this property. Moreover, intentions need not be as temporary as this argument contends. Some of our most basic intentions, like our intention to survive, can last for almost a whole life.

Still, some people complain, there is another oddity pertaining to intentions differentiating them from what we normally call dispositions. Typically, intentions do not result in actions in any immediate way, even if the premises of the syllogism are fulfilled. Suppose that Chirac intends to renew his presidency. Suppose he knows all the necessary measures to be taken to reach this goal. This nevertheless need not result in any immediate action of his. There is then no action of his right now, although all the premises of the syllogism are fulfilled right now.

The intuitive explanation, of course, is that there are no present actions which could be expedient for his purpose. Chirac believes that the actions which are necessary for attaining the presidency should be performed at times later than this time.

This feature of a delayed consequence is obviously often a feature of intentionality. But is it really peculiar to the field of determination of human action? Can we not find any "natural" dispositions that have delayed consequences in the way indicated above? In fact, we have in our example about the bomb with a delayed release mechanism, which was presented in chapter 4, answered this question. Also in that case there is a delay from the time of the occurrence of the initial condition to the time of the consequence. The feature of delay, then, is not peculiar to intentions.

Fourthly, do intentions really have all the properties that we normally ascribe to dispositions? In particular, there are those features which are sometimes called dispositional *grounds* or grounds for dispositions. But are there any similar grounds for intentions?

I have previously alluded to these features. I said that a hypothetical does not constitute a complete analysis of a dispositional concept. When one ascribes a disposition to an object one does not just ascribe a hypothetical to it. One ascribes a property to it, which is, among other things, such that a hypothetical holds true. And this property can of course, in principle, be described in categorical terms. We can describe the molecular structure of a window in categorical terms. Hence, we can independently describe that structure which accounts for the fact that the window breaks when it is hit.

Now somebody might ask: Is this really possible with intentions? Is there a property belonging to a human being which accounts for the truth of the intentional hypothetical and which is describable in categorical terms? Here we must distinguish between the questions: is there such a property? and: do we now know what it is like so that we can describe it?

I admit that we do not now know of any such property. We cannot describe any property of me in, for instance, neurophysiological terms, which will account for my intentional hypothetical to go to Prague in a few days. This is, however, true of many other dispositions as well. We do not know exactly what structure of, for instance, a particular chromosome accounts for its tendency to divide in exactly the way it does. Nevertheless we have the firm hypothesis that there is such a structure which will eventually account for the truth of this tendency.

Similarly, I find it more than plausible to assume that there is an internal structure of a human being which accounts for the fact that a certain intentional hypothetical of hers is true. I do not here want to speculate about the exact nature of this structure. I do not even have to make a decision whether it is mental or perhaps ultimately neurological. My own metaphysics, however, happens to be materialistic. I believe that there is an ultimate neurological or neurophysiological ground accounting for the intentional hypothetical.[24]

I do not however believe, and I am not forced to assume, that there has to be any particular neurological set-up. There could be a great number of alternative set-ups accounting for the same intention. As there are, of course, many alternative molecular structures which could account for brittleness.

To conclude, I find it more than plausible that intentions are like dispositions in general in the way that they have grounds which are in principle stateable in categorical terms.

And now we have reason to return to our question about causation. It is now easier to see in what way dispositions, and intentions in particular, can be viewed as causes. It is obvious that a sufficient condition of a window's breaking includes not only that it has been hit but also that it has a particular molecular structure. The molecular structure is one of the non-redundant parts of the sufficient condition of the window's breaking. Hence, it is also one of the causes. (I am now using a rather general notion of cause. I ignore the pragmatic considerations that one could adduce as a reason for picking out one particular part of a sufficient condition as *the* cause.)

---

24 In fact Hempel himself discusses the case of dispositional explanation in [36], pp. 457-463.

A sufficient condition for a human being's acting does not just include her beliefs (or their neurological counterparts) but also the ground of the intention, be it physical or mental. Hence, the intentional ground can of course be viewed as a cause of the action in question.

Does this not, then, have important consequences for the explanation of actions? Can't we then construct explanations where the dispositional or intentional grounds are included among the initial conditions? Can't these explanations be completed by a universal law connecting the initial conditions, and the consequence, whether it be a description of an action or some other event? And would this not be an ordinary deductive-nomological explanation? Have we then not gone all the way back to Hempel?

Yes, indeed, this is in principle possible, as I see it. At least, I cannot for the moment see any convincing argument why it should not be possible. The trouble, however, is that we do not have the material for giving such explanations in general. Moreover, there is little hope that we will have such explanations in the foreseeable future. We are left with the traditional intentional explanations. Still, as I have tried to show, these explanations can be put in a deductive conclusive form, in virtue of the fact that a disposition partly is of a hypothetical character.[25]

APPENDIX: TUOMELA ON ACTION EXPLANATION

Raimo Tuomela deals extensively with the matters discussed by me here in his very rich and comprehensive book *Human Action and Its Explanation* [102]. His general outlook and analytic programme are similar to mine. One of his main purposes is to put action-explanations into a more general framework of scientific explanation.

Tuomela also uses the intentional explanation discussed above (or more specifically: von Wright's version of it) as one of his platforms for analysis. Like me, Tuomela both tries to understand the intuitions about the logical conclusiveness of the practical inference, and argues that intentions are partial causes of actions. Moreover, Tuomela offers analyses of mental dispositional concepts which have obvious resemblances to my analysis of intention and also to my analysis of the concept of want.

In spite of this our conclusions appear to be rather different. In explicating the logical nature of the intentional explanation Tuomela does not make use of his dispositional analysis of mental concepts. Instead he introduces a general conceptual truth *L*, which he adds to the premises of the intentional explanation in order to make it a logically conclusive argument.

Expressed in ordinary language, *L* has the following form:

25 The idea that grounds of dispositions can function as causes and be governed by general laws is developed by Hempel [36]. Davidson in his famous essay 'Actions, Reasons and Causes' [22], simply notes that dispositions are frequently referred to as causes in ordinary causal explanations. For a recent discussion of the possibility of equating mental phenomena with neurological events, see A. Marras [53].

Whenever *A* intends to bring about *E* at *t* and considers the doing of *X* no later then *t'* necessary for this, and if normal conditions obtain, then *A* will do *X* no later than when he thinks the time *t'* has arrived.[26]

Tuomela does not in this formulation explicitly quantify actions and agents, but later on in the text he says that he could very well do that. And he explicitly calls *L* a generalisation, a conceptual generalisation. Now the intentional explanation can be said to be logically conclusive in virtue of the truth of *L*. (Observe that *L* is not, despite superficial similarities, equivalent to my intentional hypothetical. *L* cannot be a partial analysis of the concept of intention. The expression "*A* intends" is part of *L* itself.)

About this generalization *L* Tuomela goes on to say the following on p. 181:

> *L* can be regarded as representing a conceptual truth within or relative to the framework of agency. It just makes a conceptually true statement about certain (generic) intentions, beliefs and actions.

The clause "relative to the framework of agency" is important to Tuomela. He admits that *L* is not always true. There may be instances of all the premises of *L* being fulfilled but no action being performed. In this context Tuomela considers an argument called the tyrant-shooting argument.

> Suppose an agent has decided to shoot a tyrant. He thinks that it is necessary to pull the trigger of his revolver when standing in front of the tyrant. Let us assume that all of the so-called normal conditions ... are satisfied. However, when the time ... for acting comes, nothing happens. The agent does not even set himself to pull the trigger. No effects of paralysis or anything related are found — the normal conditions hold true (p. 188).

I have already indicated my standpoint concerning such a case. I would consider the absence of action a conclusive sign of the falsity of one of the premises, perhaps the intentional premise. If all normal conditions obtain and the belief-premises are really fulfilled, it cannot be true that the agent really intends to shoot the tyrant. He may want to, he may have a strong want to do so, he may even at one point have decided to do so, i.e. formed an intention to do so, but the absence of action still is a sign that at the moment of action the intention must have vanished.

Tuomela's reply is different. Since I do not completely understand his solution, I shall quote his central passage on this issue in extenso.

> My solution to the puzzling situation is as follows. While I accept the descriptive practical syllogism as logically conclusive ... I do not regard the tyrant-shooting example as a counterexample to the logical conclusiveness of the practical syllogism. Considering that example as it was described

---

26 Tuomela's conceptual generalisation is quite similar to the following law of action which has been proposed by P. Churchland [19]: If *A* wants *P* and believes *X*-ing is a way to bring about *P* and that there is no better way to bring about *P*, and *A* has no overriding want, and knows how to *X*, and is able to *X*, then *A X*'s. (The law is here slightly simplified.) Tuomela [102], pp. 287-288, unfortunately misunderstands my own analysis of the concept of intention. He mistakenly interprets my intentional disposition as a universal law. As a consequence he maintains that a further premise entailing that the agent wants *P* is missing in my schema for intentional explanation. Tuomela [104] has subsequently enlarged his theory of action to encompass the notion of collective action and other social concepts.

> earlier, I think nothing conceptual is missing in it. What is missing are the underlying dynamics: there is no causal connection between the intending and the action. That is, the agent's intending did not turn into a causally effective trying at all in this example, and this suffices to explain away the example as a counterexample. To put this differently, the agent was not fully "one of us" or a "full-fledged" member of our community at the time when the action was supposed to occur (p. 190).

It seems then that Tuomela requires some conditions to obtain about the agent before applying the intentional explanatory schema at all. The agent must at the time of application be a real agent, he must be "one of us".

It is difficult to see what this means and how much this explains. It is also difficult to see how Tuomela squares these considerations about the non-acting agent and his own dispositional analysis of wants and intentions which is given in another context in his book (see e.g. p.221).

To intend that *p*, means according to Tuomela, roughly, being in a dispositional state which is such that it will — given suitable internal and external circumstances — cause the bodily behavior needed for the satisfaction of the intention. Here, obviously the causal dynamics is part of the meaning of intention, part of the concept of intention. It is then difficult to see what it means to ascribe an intention to a man who is not one of us. Do we then not have another concept of intention?

The relationship between these pieces of analysis is thus not quite clear to me. Tuomela does not, as far as I can see, make cross-references between his dispositional analysis of wants and intentions and his analysis of the intentional explanation. (He makes such references later on in the book, but there he does not discuss the possibility of absence of causal dynamics.)

One of the results of this is the following: According to Tuomela's explicit interpretation of the practical syllogism (as he calls the explanatory argument), the syllogism is not sufficient as a scientific explanatory schema. (It may, however, suffice for the purpose of interpretative understanding.) Since, according to Tuomela, the syllogism itself does not guarantee any causal connection, a scientific explanatory schema must add further premises stating the actual causal relation between the intentions and beliefs on the one hand and the action on the other. One wonders, however, whether these additions would have been necessary had he taken his own analysis of intentions and wants more seriously. A solution very similar to mine could be drawn from the material in Tuomela's book. He does not, however, himself draw it.

CHAPTER SIX

ON THE LOGICAL FORM OF INTERACTION

INTRODUCTION

When a robber forces a bank clerk to hand over the bank's cash, we have to do with a form of what we call human interaction. An agent, the robber, acts with an intention vis-à-vis another agent, the bank clerk, so that the latter as a result thereof performs another action. However, our instance exemplifies only one of many different types of interaction. There are in everyday language, as well as in the languages of social science, a large number of names of other types of interaction. Among these are, for example, *threat, warning, persuasion, advice, recommendation* and *information*. These concepts are of central importance for the description of social reality and there is a great theoretical need that they should be analysed and related to one another.

In the following I shall limit myself to an analysis of such interaction as involves two agents, i.e. a primary agent and a secondary agent. Another limitation is that it applies only to those cases where the secondary action is an intentional action. I ignore, in other words, those cases where one agent exercises physical force upon another, for example pushing him down a steep slope or carrying him up a flight of steps. The general form of the kind of interaction I shall be dealing with is therefore the following: an agent *A* performs an action with an intention vis-à-vis another agent *B*, so that *B* as a result of this performs an intentional action.

THE ANALYSIS MODEL

As an underlying model for this analysis I shall use the intentional explanation schema presented and analysed above.

(i) *A* has the intention to realise *P* (the intention component)
(ii) *A* believes that she is in *S* (the situation component)
(iii) *A* believes that in *S* she will not bring about *P* unless she does *F* (the means component)
(iv) *A* can perform *F* (the possibility component)
―――――――――――――――――――
(v) *A* performs *F* (the action component)

A moment's reflection will show that this model may not only be used to analyse explanation of action but — and for analogous reasons — may also be used to analyse determination of action. The model's ability to explain lies precisely in the fact that the components involved determine the action that is subsequently performed. Components (i) to (iv) explain by virtue of the fact that they determine (v). (Since I shall no longer be talking about explanation but determination of action I shall use the neutral term *practical syllogism* to denote the structure (i) to (v).)

This insight can be used for the analysis of interaction in the following way: When an agent *A* influences an agent *B* to perform an action *F* intentionally, *A* then sees to it that *B* gets a complete set of determinants of the (i)-(iv) type. In the most extreme case *A* supplies *B* with a complete practical syllogism, i.e. supplies the latter with an intention and two types of belief as well as with ability and opportunity for action. But in the normal case she does much less, of course; she supplies *B* with one or at the most two components in the light of the fact that *B* already has the other components that are necessary and together sufficient for a particular action to result. Just as ordinary causation happens in such a way that a cause results in its effect given a fixed background, so one action determines another action in the light of an existing set of intentions and/or beliefs and/or ability and opportunity.

Thus, the mechanical pattern is already given. Determination can occur through all four components being supplied, or three components in four possible combinations, or two components in six possible combinations, or a separate component of the four, individually.

I shall here confine myself to elucidating and classifying some of our most common interaction concepts in the light of this model. I shall divide the following part of the study into two sections. The first deals with perhaps the most common forms of interaction that have to do with operations on the second and third components in the intentional explanans. The second section deals with operations both on the fourth and on the first component; the latter being from the philosophical point of view the most problematical component.

ON THE MANIPULATION OF THE SECOND AND THIRD COMPONENTS IN THE EXPLANATORY SCHEMA

The first type of interaction I wish to study in detail is the following: The secondary agent already has an intention and is induced to act in a particular direction on the basis of this intention through being supplied by the primary agent with components of type (2) and/or type (3). To see what this interaction can involve, let us look at the following example.

The Minister of Health for Scotland, the Tory Mr John Smith, has the intention to stay in power. We can assume that, for the moment and in the near future, this intention of Mr Smith's is not operative, i.e. that he does not do anything in particular to stay in power. It would be entirely possible, however, to get him to act on the basis of this intention, i.e. the intention can constitute the basis for interaction. Consider the following case:

> The MPs of the Scottish Nationalist Party threaten to vote with the Labour Party in order to get much stricter alcohol regulations in Scotland.

Assume now that the parliamentary situation is such that Smith's position is actually threatened. If the SNP votes with Labour, Smith's liberal alcohol policy cannot be pursued. Since he has invested much prestige in this issue he would have to resign.

What happens from a formal point of view in this case is that the MPs of the SNP create, or threaten to create, a situation in which Smith must act if he wishes to stay in power. In other words, the situation around Smith has changed. Smith becomes aware of this fact and finds that in the light of this new situation he must make certain concessions to the MPs of the SNP and accept a compromise concerning the alcohol policy.

The interaction consists in this instance of the following: The MPs see to it that Smith compromises on the alcohol issue by changing the situation and thereby changing component (2) in Smith's practical syllogism. We can, as it were, assume component (1); component (4) already exists. Component (3) has in this case been added through Smith's having himself calculated, on the basis of his reading of the situation, the means necessary for him to stay in power.

In ordinary interaction terminology we can describe this situation using at least two terms, one of them more specific and the other more generic. The specific term has already been used: The MPs get Smith to compromise by means of threat. (This may perhaps also be described as a case of force or coercion. For a more complete analysis of the latter concepts see below.)

Against the same background as above, i.e. Smith's intention to stay in power, we could conceive other forms of interaction. Assume that the Secretary of the Conservative Party were to inform Smith about a coming threat from the MPs. In this case Smith learns about the situation in a special verbal way. This form of interaction, i.e. information about a threat, we usually call warning. We say that the Secretary gets Smith to compromise on the alcohol issue by warning him.

This is still an operation on component (2) but in a completely different way. The Secretary does not himself change the situation; he merely informs about it.

It is, of course, important to observe that the two forms of interaction which we have just studied are not mutually exclusive. The Secretary's warning does not exclude the SNP's threat, and vice versa. Both actions could very well occur in one and the same interaction situation. The relation should perhaps be so described as to place the SNP's threat first in the causal chain, with the Secretary's warning as the intermediate stage and Smith's insight regarding the threat as the final stage.

A warning is often associated with advice or recommendation. Perhaps the Secretary does not have confidence in Smith's judgement in the situation that has arisen. In his opinion concessions should perhaps not just be made to the SNP but also to the health organizations, in order to avoid suspicion that Smith is just being opportunistic on this particular occasion.

If the Secretary also succeeds in convincing Smith as to the proposed line of action, then he has interacted in the double sense of his having directly induced both the components (2) and (3) in Smith's practical syllogism. The action that is performed in this case is consequently the more complicated action inasmuch as Smith compromises with more than one party.

ON INTERACTION BY THE MANIPULATION OF THE FOURTH AND THE FIRST COMPONENT

So much for interaction with respect to components (2) and (3) in the practical syllogism. Regarding component (4) I can be brief. The reason is that I have already in a sense dealt with it. It is the case where one influences people simply by making it possible for them to act in a desired direction.

Let me here just add a formal distinction. Enabling a person *A* to perform an action *F* can mean two different but complementary things. On the one hand one can induce in *A* the internal ability to perform *F* (given a set of circumstances); on the other hand one can see to it that the right set of circumstances obtain, i.e. *give A an opportunity* to do *F*. Direct manipulation — medical or otherwise — of a person's body or mind in the name of health promotion, for instance, is aimed at giving a person the internal ability to act in a certain direction. To manipulate the external environment in order to make it possible for a person to do what she wishes is to give her the opportunity for action.

The enabling conditions in the case of Smith's compromise with other politicians or officials are certainly very trivial. He may simply be given the opportunity to meet or otherwise contact the relevant parties.

The study of factors influencing *the first component* in a practical syllogism, however, presents us with interesting and deep philosophical questions.

How can one cause a person to acquire a new intention? We can very well ask the question since it is so natural for us to regard interactive influence as something which occurs in the light of an already existing will or intention. As far as I can see there are two important instances to distinguish:

(1) determination of an intention in the light of another more general intention,
(2) determination of a fundamentally new intention which cannot be derived from another intention of a higher order that the agent already has.

The first case can be illustrated by the following example:

> The Secretary of the Conservative Party in Scotland made Smith decide to go in for a comprehensive health-promotive programme by convincing him that this would lead in the long run to an improvement in public health and a decrease in crime.

When we analyse this example we find that it is not a case of genuine creation of an intention. The Secretary gets Smith to decide on something, but only in a secondary sense. He demonstrates that a certain goal is a necessary subgoal for the realisation of something that Smith already wants and has the intention to pursue. We can describe the situation schematically in the following way:

(1) Smith has the intention to reduce crime in Scotland,
(2) Smith believes that he is in a situation *S*,
(3) Smith is convinced by the Secretary that health promotion in the long run, given situation *S*, contributes to the reduction of alcohol abuse and thus to the reduction of crime.

---

(4) Smith acquires the derived intention to go in for health promotion.

Thus, strictly speaking, we find that what the Secretary brings about is not component (1) but component (3) in a schema of a higher order. As a consequence of this there then arises a new intention of a lower order.

This process of a new intention arising in the light of an already existing intention of a higher order illustrates the theoretical importance of noticing *intention transmission* of the kind more fully analysed in chapter 4. It is obviously a common form of determination of intentions.

The deep philosophical question remains. Can wills or intentions be created by interaction in a different way? Can one agent induce a genuinely new intention in another agent? It could be plausible to say that religious conversion or chemical interaction, the latter for instance through injection of a narcotic drug into a person, are such cases. I shall however not here attempt to explore these cases in order to see if they are genuinely different cases.

Let me summarise this analysis by means of the following schema. An agent *A* can influence another agent *B* by doing one or more of the following:

by showing *B* that she must decide to realise *P* given her higher-order intention to attain or maintain *Q*,

by creating a situation *S* which gives *B* a reason for doing *F* in the light of *B*'s intention to realise *P*,

by informing *B* about the situation *S* (this may be a case of *warning*) which *B* sees as a reason for doing *F* in the light of *B*'s intention to realise *P*,

by informing *B* of the appropriateness or even necessity of doing *F* in *S* in the light of *B*'s intention to realise *P* (this is then a case of advice or recommendation),

by enabling *B* to do *F*, e.g. by means of medical care, or by giving *B* an opportunity to do *F*.

Given this schema it is easier to say something about the act of *persuasion*. I suggest the following analysis of *A* persuades *B* to do *F*.

> *A* intentionally sees to it that *B* does *F*, by rationally influencing any subset of components (i) to (iii) in a syllogism of *B*'s that has doing *F* as a conclusion.

Two comments. First, it is important that the influence is rational in the sense of being purely epistemic. The means should be a transmission of information solely, not any other change in the world. Bringing about a change in a situation and thereby indirectly changing a person's belief about the situation is not persuasion. Nor is enabling a person to act or giving a person an opportunity for action a case of persuasion. Persuasion need not, however, be verbal. Other means of communication than the conventionally linguistic one are possible.

Second, persuasion is different from warning, advising and recommending in yet another way. The verb "persuade" is a success-verb, which presupposes that the secondary agent actually performs the action in question. The other acts of influence may fail in this crucial respect.

Let me now particularly consider the set of interactions denoted by the term *force* or *coercion*. I shall then as a paradigm case use the one which introduced this chapter: A robber forces a bank clerk to hand over the bank's cash. Assume also that in doing so the robber threatens the clerk with a gun, i.e. threatens to kill the clerk unless the latter obeys his orders. There is then a collateral relation which could be expressed thus: The robber forces the clerk to hand over the cash by threatening her life.

The first observation to be made is that at least this paradigm case of force involves an influencing of the second component of the syllogism. The robber indeed creates a change in the background situation for the clerk and he also sees to it that the clerk becomes aware of this change. That is, the robber both threatens and warns. But which then is the practical syllogism in accordance with which the clerk acts when she hands over the cash? A plausible reconstruction is the following:

(i) The clerk has the standing intention to survive and remain unhurt;
(ii) The clerk considers the new situation where her life is threatened;
(iii) The clerk believes that she cannot guarantee her survival unless she obeys the robber's orders.

Since he is unprevented and capable the clerk follows the instructions of the robber.

So far this interaction has been explicated according to the general model. The clerk's action can be explained as a case of interaction where the second component has been influenced. But what, then, distinguishes force or coercion from other kinds of interaction where the secondary agent has been given a reason for action?

A typical premise (but perhaps not necessary for all uses of the terms) is that the secondary agent normally has no intention whatsoever to do what she is forced to do on a particular occasion. She may vehemently dislike doing F, for instance because doing F is highly immoral or even criminal. The second and most crucial premise is often put in the following terms: B cannot on this occasion avoid doing F, or doing F is irresistible for B.[27]

I shall here argue that the second premise must be analysed carefully. The wordings "B cannot avoid doing F" and "Doing F is irresistible for B" are very unhelpful and I shall — in accordance with the analysis given already in chapter 4 — suggest a very

27 This particular phrasing can be found in R.R. Faden and T.L. Beauchamp [29], p.339.

different analysis. Consider the case of the bank clerk. First, there is no disability whatsoever on her part. What is the sense of saying that she cannot avoid handing over the cash? Certainly, she has the ability to abstain from handing over the cash and she also has the opportunity to abstain. (See my previous analysis of the relevant notions.) Indeed, we know of many clerks who as a matter of fact have refused to hand over the cash, and this has not proved difficult in pure capacity terms.

Thus in most instances it is clear that the clerk can very well avoid handing over the money and there is no irresistibility about the situation. There is, however, another very important factor. The clerk has a very firm intention, viz. the intention to survive. Moreover, this is an intention that she is not prepared to relinquish for any reason whatsoever. In the situation under discussion the clerk experiences a conflict. She wants, and has certainly also formed the intention, to guard the bank's money. She thinks it wrong to hand it over. On the other hand in the conflict with her intention to survive, the former intention has to be relinquished. She decides to act contrary to her normal duty. Thus in the light of an intention that the agent is not prepared to compromise on, there is only one act to perform. Under such circumstances the agent acts under force.

So far I have talked about force as an interactive term, i.e. where a primary agent creates a compelling situation. Compelling situations may, however, also arise naturally and can be given a parallel analysis. Some such instances of force will be analysed in chapter 14, on mental illness and forensic psychiatry.[28]

ON INTERACTION IN TERMS OF PREVENTION OF ACTION

So far I have exclusively discussed various forms of positive interaction. But the case of someone preventing someone else from performing a certain action is certainly also a common variant of interaction.

The practical syllogism is obviously very useful also for this purpose. A primary agent can prevent a secondary agent from performing *F* by eliminating one or more of the components in *F*'s practical syllogism. Consider an example:

John prevented Peter from walking down the street to the shop by

- (i) convincing him that he should give up his intention to buy food
- (ii) reminding him of the dangerous dog he might encounter
- (iii) convincing him that this was not the quickest way to the shop
- (iv) confronting him with obstacles.

Observe that these are just examples of preventive measures classified by the schema. In position (iv), for instance, we could place various kinds of manipulation of the

28 The analysis presented here is strictly theoretical. I choose such an analysis well aware of the fact that some of the interactive concepts are often used with an evaluative (moral) load built into them. A concept such as "manipulation" is often defined as a morally improper way of interacting with another human being. Sometimes "coercion" and even "persuasion" are also thus defined. See for instance the discussion in R.R. Faden and T. L. Beauchamp [29], pp. 337-373, concerning manipulation and coercion. In my own view it is important, first, to theoretically analyse the various modes of interaction. The result of this analysis can then be used as a platform for an ethical investigation.

agent's body and environment which could have a preventive result. See my analysis below of the notion of practical possibility.

We can thus see that there is an important asymmetry between determination and prevention. Whereas determination presupposes a complete set of determinants for the action in question, prevention just presupposes the removal or the non-existence of one of the determinants of a complete set. (We assume then of course that there is no other sufficient determinant present.)

There is an important distinction to be made between *two* kinds of *action prevention*. The first case can be illustrated by the following example: Peter *is about to* walk down to the shop. John changes Peter's mind by removing his general intention to go to the shop. (This in turn can be done in various ways, for instance by convincing Peter that he has enough food in the refrigerator.) As a result Peter decides not to go to the shop. And in so deciding he intentionally remains at home. Here, the prevention of an action can very obviously be analysed as causation of omission. A very different kind of prevention takes place when there is no change of mind on the part of the secondary agent. There is perhaps just a risk that the latter will acquire a complete intentional set-up for the performance of a certain action. The primary agent prevents this becoming the case by preventing one of the components from becoming materialised.

The first case can be analysed in terms of two *Do*-operators. The most natural analysis would be: *Do (A) Cause (A) Do (B) E*, i.e. *A* causes *B* to omit to perform an action *F*, which could be translated into, *A* causes *B* to let a certain episode be the case.

The latter case can be illustrated by a primary agent who sets up obstacles or otherwise disables a person; Peter is caused not to act in the simple sense that he is caused to remain where he is. The whole episode thus turns out to be just a causative state with only one active person involved. The formula would be: *Do (A) Cause (A) (Remain s).*

CHAPTER SEVEN

TOWARDS A THEORY OF ABILITY AND DISABILITY

INTRODUCTION

So far actions have been analysed from the point of view of definition, explanation and interaction, and as elements in the theory of episodes. I shall now turn to a more thoroughgoing analysis of component 4 in the practical syllogism, viz. what in the following will be called a person's practical possibility of performing an action.[29]

ON THE POSSIBILITY OF PERFORMING AN ACTION

What does it mean, more precisely, to say that a human being can perform an action? The term "can" is highly ambiguous; it admits of such diverse interpretations as (a) logical possibility (nine can be divided by three) (b) epistemic possibility (for all I know, he can be thirty years of age) (c) physical possibility (people can survive only with oxygen) (d) ability (John can speak Russian) (e) authority (this university can issue Ph.D. degrees) (f) opportunity (Peter can cross the road now).

If we limit ourselves to human beings and their relations to actions, we could take at least the last three interpretations into account; when we say that *A* can perform *F* we might mean either *A* has the ability to perform *F*, *A* has the authority to perform *F*, or *A* has the opportunity to perform *F* — or some combination of these. For the purposes of the present discussion it is particularly important to distinguish between a person's ability and her opportunity to perform *F*. (For many theoretical purposes authority can be viewed as a conventional circumstance, as a kind of opportunity.)

When a person has both the ability and the opportunity (including authority) to perform a particular action, then she can perform it in the strongest sense of the word. This strong sense will here be called *practical possibility*.

How should practical possibility be characterised? There are at least two traditional routes for such a definition. On the first and best-known route possibility concepts are defined conditionally, in analogy with dispositional concepts. For example: "It is practically possible for *A* to swim" means that, if *A* tries to swim, then *A* succeeds in swimming. On the second route, possibility concepts are defined in terms of possible-world semantics: "*A* has the ability to swim" then simply means: in some worlds (or in some circumstances) *A* swims.

For the purposes of this investigation I shall not take a stand concerning the ultimate choice of theoretical definition. I shall not propose a definition of practical possibility. It suffices here to provide a test for the application of the concept. This test is founded on the traditional analysis. We can ascertain whether it is practically possible for a per-

29 In this chapter I essentially reproduce my analysis in *On the Nature of Health* [71], pp. 41-57.

son to do *F*, by letting her *try* to do *F*. From a test where *A* tries to *F* and succeeds in *F*-ing, we shall conclude that it is practically possible for *A* to do *F*.[30]

Consider now the conditions for a person's practical possibility of acting. These conditions differ, of course, depending on the type of action and whether we are talking about ability or opportunity. I shall first discuss conditions for *ability* and such conditions as hold for all kinds of action, in particular for *basic actions*.

I shall first take into account the fact that actions are intended. A general ability to perform an action thus presupposes an ability to form the intention to perform the action in question. This indicates that certain mental preparation is necessary for a person to act.

It is impossible for *A* to intend to perform *F* if she is completely unaware of *F*. This kind of situation is not so common with basic actions as with many generated actions. (See the discussion below.) Still, the point is also relevant for basic actions. People are not aware of all the possible movements they can make with their limbs. Hence, there are certain movements they will never intend to make.

Also, there are mental factors which may prevent a person from performing a particular action. She may find an action so revolting that she would never intend to perform it. This again is more common with certain complex actions; most unethical actions, for instance, are complex actions.

A further interesting case of mental prevention is the one where an agent is continually *convinced that she is not physically able* to perform the action in question. If this is so she will never form the intention to perform it.

Thus factors such as ignorance of an action, revulsion at it, or conviction of one's physical inability to perform it, will prevent the realisation of the first stage in acting, i.e. deciding or intending to act and setting about acting. These factors, which are not generally acknowledged in the context of ability, have particular importance for the theory of health. Many types of mental illness can be located in defects among the antecedents of intending.

For the realisation of the second stage of action, its actual performance, there are certain obvious requirements. With basic actions these requirements all concern the

---

30 R.R. Ayers [6] argues strongly, in particular on pp. 125-144, against a conditional analysis of power-concepts in general. A central argument is the following:

> No kind of possibility-statement can conceivably be analysed into hypotheticals like those suggested, simply because, as we have previously remarked, *p* entails that *p* is possible whereas *p* does not entail that if *p* then *q* (p. 128).

Ayers is particularly suspicious of analysing human ability in terms of a conditional where trying is the antecedent. Trying, he says, is no event or action antecedent to the "real" action. Trying is identical with the action tried. It is only that some instances of trying are abortive actions (i.e. when trying results in failure). A. Kenny [44], on the other hand, disputes the general validity of the inference: *A* does *F* entails that *A* can do *F*. Thus he disputes Ayers' main argument against conditional analyses of the concept of ability.

> A hopeless darts player may, once in a lifetime, hit the bull, but be unable to repeat the perfomance because he does not have the ability to hit the bull...the distinction between luck and skill is not a marginal matter in this context: it is precisely what we are interested in when our concern is ability, as opposed to logical possibility or opportunity. (p. 136.)

For an excellent recent treatment of ability- and power concepts, see P. Morriss [59].

biological make-up of the agent. This make-up can be divided into various aspects. One is that the agent's neurological machinery is in order. She must not, for instance, be paralysed. Other aspects involve such things as the muscle tissues being sufficiently developed, the joints functioning properly, and so on.

The condition of opportunity is easy to characterise in the case of basic actions. Here opportunity consists merely in the *non-existence of external preventive factors*. A man *A* has the opportunity to raise his hand if nothing physically prevents him from doing so.

These, then, are the background conditions for the practical possibility of performing basic actions. Consider now the complex actions *accomplishments* and *projects*.

By definition, the performance of an accomplishment requires the performance of some basic action. A second requirement is that the accomplishment can *in fact* be generated. (Note that this may depend partly on the agent in question.) A third requirement is that the agent knows that there is a situation which constitutes the opportunity to generate the accomplishment in question. This entails either that she has some causal knowledge, i.e. knows what happens, given a particular basic action in a particular situation, or that she has some conventional knowledge, i.e. knows of a particular action-generating rule and what it says about the required circumstances. (In some cases both kinds of knowledge may be presupposed.)

I shall now collect these requirements (together with the ones noted above) into one schema. The following symbols will be used: *Acc* for accomplishment, *Pro* for project, *B* for basic action, *O* for opportunity and *S* for action-sequence.

It is practically possible for *A* to perform an accomplishment *Acc,* if and only if,

(i) there is at least one action-chain, *B ...Acc*, given an opportunity *O*;
(ii) *A* believes that (i), feels no revulsion against performing *Acc*, and believes that she is physically able to perform *Acc*;
(iii) it is practically possible for *A* to perform *B*;
(iv) *O* is present;
(v) *A* identifies *O*.

Let me now turn to the case of projects. The practical possibility of carrying out a project must involve the practical possibility of performing each action which is a member of some action-sequence constituting the project. (As I have said, there are often alternative ways of carrying out the project.)

But the practical possibility of performing each member of a set of accomplishments does not suffice for the performance of the project. Again the agent must have a considerable amount of knowledge. I can summarise the items that she must know:

(i) *A* must be aware of at least one action-sequence constituting the project.
(ii) *A* must know what constitutes the opportunity for all members of this sequence.
(iii) *A* must know how these opportunities are to be identified.

A further important element in the performance of some projects is the element of *co-ordination*. It is sometimes required that one cannot only perform each of the basic actions or accomplishments involved in the project, but also coordinate them into a sequence with special properties (for instance, properties of time, force or elegance). For instance, to produce a melody it is clearly not enough to produce the right notes, one at a time. The components of the tune must be coordinated in a particular way for the result to be music.

Let me express the requirements for carrying out a project in the proposed formal manner:

It is practically possible for *A* to carry out the project *Pro,* if and only if

(i) there is at least one action-sequence *S*: *Acc1...Accn* constituting *Pro;*
(ii) *A* believes that (i), feels no revulsion against carrying out *Pro* and believes that she is physically able to carry out *Pro;*
(iii) it is practically possible for *A* to perform each of *Acc1...Accn*, given the respective opportunities;
(iv) *A* is able to coordinate each of *Acc1...Accn* in the appropriate way;
(v) the required opportunities actually arise;

and

(vi) *A* identifies these opportunities.

ON NOT BEING ABLE TO ACT

My analysis of the notion of practical possibility of action provides me with the tools for seeing the variety of reasons why people are sometimes not able to perform the actions they want to. I shall sum up below the nature of the factors which can prevent the realisation of an action. I shall thereby use the concepts and distinctions previously introduced in the book.

Non-ability to perform a basic action, say the movement of one's leg, can be due to a neurophysiological disorder, immediate external prevention or a serious mental disorder, entailing weakness of will. This, then, is a rather limited set of factors.

Non-ability to perform an accomplishment *Acc* can be seen to depend on not less than the following factors:

(i) There is no action-chain *X* → ... *Acc*, where *X* is a variable for basic actions;
(ii) *A* is not aware of any action-chain *X* → ... *Acc*.

On the supposition that there is an action-chain *B* → ... *Acc* known to *A*, we have the following possibilities:

(iii) *A* is not able to perform *B*;
(iv) *A* does not know about her ability to perform *B*;
(v) There is no opportunity for *Acc* by the performance of *B*;
(vi) *A* does not identify any opportunity for *Acc*.

Let me illustrate some of these reasons for non-ability by considering the accomplishment of starting the engine of a car. *A* may be unable to start the car for the reason that

(i) The engine has broken down. There is no mechanism being initiated by the starting button;
(ii) *A* does not know of any way of starting the engine;
(iii) *A* has broken her right hand and cannot put her finger on the starting button;
(iv) *A* does not know that she can move her hand properly (she is, for instance, not aware that her hand has healed after having been broken);
(v) *A* is not in the position to press the button: she is at some other place and hence there is no opportunity for starting the car;
(vi) Although *A* is sitting in the car she cannot find the button.

Consider a further illustration, where the action is a conventional accomplishment, viz. voting in a referendum.

(i) There is no referendum; it has been cancelled;
(ii) *A* does not have the authority to vote: she has not yet acquired the relevant citizenship;
(iii) *A* does not know about the referendum or about the conditions for taking part in the referendum;
(iv) *A* has broken her hand; hence she is unable to put the voting envelope into the urn;
(v) *A* does not know that she is able to move her hand;
(vi) *A* is not in the position to vote: she is abroad;
(vii) *A* does not know that she happens to be in the room where the voting is taking place.

The ways in which an agent may be unable to carry out a project are then easy to see. I shall not elaborate further on them here.

ON ACCOMPLISHMENTS AND PROJECTS REQUIRING MORE THAN ONE AGENT

Many projects in particular require the cooperation of a number of agents. The construction of buildings, the production of TV-programmes and the ruling of countries are such examples. It is almost a necessity that people representing various kinds of competence come together in order to succeed in such complex enterprises.

I shall not here in general enter into the very complex analysis of collective action and its conditions. I shall only notice one interesting special case which is of some importance in my discussion concerning autonomy below. This is the case where a single agent is said to perform an action, but where at least important parts of this action are "physically" performed by someone else. The paradigm cases of this are certain actions performed by heads of states or heads of companies. Consider the minister who is said to write a bill on the defence policy of a state. In actual fact she probably does not write a word of this bill. That is done by a number of her secretaries, given some pieces of advice and some general instructions by the minister. Thus when we reconstruct such actions of authority we shall have to acknowledge the following: the sequence of actions constituting a project such as the writing of a government bill consists partly of subprojects and accomplishments actually performed by other people than the main agent.

ON ABILITY AND DISABILITY

The list of factors necessary for the realisation of actions is long. In particular on the accomplishment and project level the list becomes extremely complicated. Thus when it is not practically possible for a person to perform a particular action this can be due to a great number of factors.

From the point of view of *human welfare* all these factors are of importance. If a person finds the performing of an action important for realising a goal she has, then it is important that she knows about and tries to handle all kinds of factors which can prevent her agency. Thus all preventive factors are relevant in the general theory of welfare.

For that part of the theory of welfare which deals with health we may narrow the scope somewhat. We can, in very rough terms, leave out the opportunity aspect of practical possibility and focus on the ability aspect. This means: are the person's internal — bodily and mental — resources for performing the health-defining actions sufficient?

Thus my interest in the undermining of action will from now on be focused on disabling factors — factors which are internal to the agent's body and mind. Let me first rehearse which types of factors these are. I shall consider the different action categories in turn.

Basic actions: Lack of bodily or mental integrity; ignorance of the action-type; belief in the impossibility of performing the action; revulsion against the action-type.

Accomplishments: Lack of bodily or mental integrity for performing the involved basic action; ignorance of the action-chain constituting the accomplishment; belief in the impossibility of performing the accomplishment; revulsion against the accomplishment; non-identification of opportunity for action.

Projects: Lack of bodily or mental integrity for performing the involved basic actions in the way required; ignorance of the action sequence constituting the project; belief in the impossibility of carrying out the project; revulsion against the project or some of its necessary components; non-identification of opportunity for action.

I shall now introduce a further dimension to the analysis of ability. Let me first note that the practical possibility of performing a particular action is dependent on how the action is specified, in particular regarding place and time. It may be practically possible for *A* to perform *F* in a particular situation *S*, but not in *S1*. The reason for this may be that *S* provides the right opportunity for *A*'s performing *F*. For instance, it is now practically possible in Sweden for a foreigner, who meets certain conditions, to vote in local government elections. A few years ago this was not practically possible; and it is not practically possible in several other countries.

To say that it is practically possible for a particular agent *A* to vote in the local elections means, according to my suggested test procedure, that she would actually succeed in voting if she tried. But what is the corresponding relation regarding *ability* in a particular situation? The presence of ability cannot be as easily determined as that of practical possibility. "*A* is able to do *F*" does not imply that *A* actually does *F* if she tries. *A* may be able to do *F* but still fail to do *F* when she tries because there is no opportunity. From "*A* is able to do *F*" we can merely conclude that *A* does *F* if she tries and if the right opportunity obtains.

But this purely formal solution does not square completely with our intuitions. Moreover, it does not suffice for defining a basic notion of ability for the discourse on health, for it implies that almost everybody would be able to do almost everything. Given a sufficient manipulation of opportunities, a person whom we would normally call disabled with respect to a certain action would become able. Consider the following example.

A schoolboy is taken into the cockpit of an aeroplane. He is instructed in detail by the pilot how to handle the instruments, indeed he is sometimes physically helped in the manipulation of them. Thus given the extraordinary opportunities created by the pilot, the schoolboy is able to fly the aeroplane.

But this kind of situation is not what we have in mind when we say that a schoolboy is able to fly the aeroplane. We do not mean that he can succeed in doing so, given extremely advantageous circumstances, but rather that he can succeed in doing so given an accepted ("normal") set of circumstances.

The distinction between extraordinary and accepted circumstances is obviously crucial in the theory of *medical disabilities and handicaps*. We would certainly describe a person who has lost both legs as disabled with respect to the action of moving about. But if she is sitting in a wheelchair, this is not true. Thus in order to describe the person as disabled in the first place we have to disregard the wheelchair.

But which are the circumstances to be accepted? Is it up to anyone to characterise this set? In such a case will there be any agreement between different users of the language?

I shall here propose the following analysis. There is a primary use of the locution "*A* is disabled" and a secondary one. According to the primary use, it is indeed true that the speaker herself defines the set of accepted circumstances. She in fact claims that, given what she finds to be a reasonable set of circumstances, *A* is disabled. Thus the expression "*A* is disabled" has, in this case, a clearly evaluative element. Moreover, it is a use that may fail to be communicative concerning the descriptive part of the message. Unless the speaker informs the listener of what is the accepted set of circumstances, the listener may fail to understand the message.[31]

---

31 The distinction between a primary and secondary use of the locution "A is disabled" did not exist in *On the Nature of Health* [71]. Moreover, the idea of characterising "standard circumstances" as a special case of "accepted circumstances" is a novelty in this presentation. József Kóvacs [46] has inspired me to think further on these aspects.

According to the secondary use, the speaker implicitly refers to *some commonly accepted background* — what might be called *standard circumstances*. The speaker does not herself take a stand on the normative issue concerning what is a reasonable background situation. She instead refers to what other people, the politicians, the experts or most other laymen, have accepted as a reasonable background situation for the ascription of ability and disability.

But even given this reference to some commonly accepted background situation we are left with a situation of relativity. What are counted as standard circumstances vary from epoch to epoch, and from society to society. Consider the differences in natural environment between Greenland and the Congo, or the cultural differences between life in the United States and life among the aboriginal tribes of Australia. What are counted as standard circumstances must vary enormously between these places. There may certainly be differences also within a very homogeneous society.

These observations provide an important insight into the logic of ability and disability. Ascribing an ability to a person *A* with regard to some action of hers is at the same time to describe part of the world in which *A* resides. If we ascribe an ability to *A*, then we presuppose that the standard circumstances in the society and the natural environment in which *A* lives, provide the opportunities for *A* to execute her ability. When we say that *A* is able to play tennis, we must presuppose that she lives or has lived in a society where the game of tennis is recognised. When we say that she is able to repair a car, we must assume that she lives or has lived in a modern industrialised society, etc.

But let us take a particular society at a particular time. Is the prevailing set of circumstances clearly defined, and how can it be determined? It is not, for instance, identical with (although it may partly be influenced by) what is a statistically normal set of circumstances in a particular society. There may also be big discrepancies between the sets envisaged by different politicians, health-care workers and laymen. The logic of my analysis is just the following. When the circumstance-position is left out in a locution of the kind "*A* is disabled" (and the locution is supposed to be descriptive) then there is *some* set of circumstances presupposed.

How do we determine the standard set of circumstances in this case? There is no algorithm for solving this problem. It is not a question of simple statistics, although how often certain circumstances arise certainly plays a role. What counts as an unusually advantageous or an unusually difficult situation with regard to a particular action is basically a normative question. We follow a convention adopted in our society when we say, for instance, that a person who is able to read should not need continuous instruction, or that a person who is able to walk should withstand a moderate gale of, say, 20 m/sec. A person who fails to walk in a hurricane, however, even though she tries, may still be said to be able to walk.

I conclude that the concept of ability, as normally understood, is related to a set of *accepted* circumstances. This set may be a completely personal one, i.e. a set considered reasonable by the user of the notion. Or it may be a set of commonly accepted circumstances, what I here call *standard* circumstances. Such a set of standard circumstances is relative to some natural and cultural environment. The concept of standard circumstance is moreover a normative one. The common acceptance of standard circumstances for a particular action type is partly influenced by statistical factors, but is mainly determined by the society's profile of goals.

ON FIRST- AND SECOND-ORDER ABILITY

We may talk of ability of different orders. I shall here distinguish between first- and second-order ability. So far, both my discussion and my characterisations have concerned first-order ability only. The notion of second order-ability will be defined as follows:

> *A* has a second-order ability with regard to an action *F*, if and only if *A* has the first-order ability to pursue a training programme after the completion of which *A* will have the first-order ability to do *F*.

Second-order ability is thus compatible with first-order disability, while the reverse does not hold. A person may lack the first-order ability to earn her living in Sweden. She may, however, have the second-order ability to do so. She may be able to train herself to make a good living in that country.

Note that the action of *training* must be given the same analysis as other kinds of action. When we ascribe to someone the first-order ability to follow a particular training programme, we must, as in the general case above, presuppose a set of accepted circumstances. We must moreover presuppose that it persists throughout the training process.

Thus a person who enters on a training programme, but in the end fails to acquire the desired first-order ability, need not lack second-order ability. First, the training programme may have been poor. This might indicate that the accepted circumstances did not obtain. Second, the subject may, after a while, no longer have intended to pursue the training in a proper way. This being so is still consistent with her having the second-order ability. Second-order ability may not turn into first-order ability, if the agent does not consistently try to acquire the first-order ability.

Consider now the following case. The subject is afforded adequate training facilities, and she tries to learn through the whole period. Nevertheless, after this period she still does not know how to perform the desired action. This indicates that the subject does not have the second-order ability, at least not through the whole period of training, to perform the action. We could then also say that she is *genuinely disabled* with respect to performing the action in question.

To summarise: A person has a second-order disability with regard to an action *F*, if and only if she is disabled, given the accepted circumstances, with regard to consistently pursuing a training programme to acquire a first-order ability to perform *F*.

The notion of second-order ability brings us closer to the biologically founded capabilities of man. Nevertheless, it does not and cannot completely free us from the relativity of an action to an environment. To say that *A* has a first-order ability to follow a training programme successfully, presupposes a set of accepted circumstances. It may be conceivable that certain people who lack a first-order ability, if they were put into extremely advanced and extremely expensive training programmes could achieve the first-order ability desired. But if such programmes have not been offered, or if they have not even been designed, they cannot be taken into account in ascribing second-order ability to people.

These observations have a great impact in my analysis of the medical notion of disability and handicap, as well as in my analysis of health. The analyses are to be presented in the second part of this book.

PART II

Action-Theory as a Basis for the Theories of Health and Welfare

ON THE NOTION OF HEALTH AS ABILITY

INTRODUCTION

Medical concepts, such as health, illness and disease, have been the focus of studies in the disciplines of anthropology, sociology and philosophy for a number of decades. There has been debate concerning mainly the respective merits of a medical view of these concepts, stressing the biological basis, and a sociological/anthropological view, which emphasises their dependence on the cultural and social context in which they are used. Most debaters, in particular the representatives of the social sciences, have sided with the sociological interpretation of these concepts. In the philosophical arena, on the other hand, quite different theories have been proposed; a forceful analysis essentially defending the medical view has been given by C. Boorse [9], [10] and [11], whilst many equally forceful criticisms of this view have been presented by a number of later theorists (for instance G.J. Agich [1], H.T. Engelhardt [28], K.W.M. Fulford [30] and myself [71]; for a recent response by Boorse, see [12]).

The question can now be asked: is this an important discussion? And what are its consequences for the institution of health care or for other organisations and activities in society? In order to be able to discuss these issues let me formulate the main conceptual positions as clearly as possible.

THE TWO CONCEPTS OF HEALTH

I shall first sketch a version of the medical — or what I shall call a biostatistical — theory of health. The fundamental idea here is the following. Biological evolution has created a number of species of which the human being is one. This evolution has entailed certain changes, but it has also involved a fixation and conservation of the fittest species and their specific characteristics. The human being has proved to be such a fit species.

Humans have a number of specific characteristics in terms of both structure and function. We have some idea of how these characteristics have contributed to the survival of individuals and the species as a whole. Through this knowledge we can picture a pattern for the life of a surviving individual. Through this vision we have laid the foundation for a biologically oriented theory of health and illness. A healthy human being, such a theory says, functions according to the pattern which is typical for the species man. A human being is unhealthy, on the other hand, if one or more of her functions deviates subnormally from this pattern.

A clear and well-organised theory of this kind has been presented by the American philosopher Christopher Boorse [11]. He develops his theory by first introducing the concept of disease. He defines this concept in the following way: a disease is a state of an individual which interferes with (or prevents) the normal function of some organ or system of organs of the bearer of the state. The normal function of an organ is, in its turn, interpreted as this organ's species-typical contribution to the biological goals of the survival of the individual and the survival of the species. Health can then be defined in the following simple way: *A* is completely healthy, if and only if *A* lacks all diseases, i.e. if and only if all of her bodily or mental functions fall within normal intervals.

But what is the procedure, according to Boorse, for determining the normality of functions of the species of man? According to Boorse a statistical method is the only reasonable one. The species-typical pattern is the statistically typical pattern. The species-typical cardiac frequency is that interval within which the cardiac frequency of most people is to be found. The general philosophy that Boorse advocates here is thereby very similar to the one used by clinical physiologists and clinical chemists when they try to determine normal or reference values for various human bodily functions.

The concept of disease defined by Boorse does not cover all ordinary usages of the term. A disease in Boorse's technical sense can be the case without the bearer being aware of it. It need not entail any disability and it need not call for any treatment. In order to capture these intuitions Boorse introduces the notion of illness. His precise definition of illness is the following: A person is ill, if and only if she has a disease which is serious enough to be disabling her, and which is therefore unwanted by the person and requires treatment [9].

It is important to notice, however, that illness, as defined here, presupposes the existence of a disease (in the biostatistical sense). Disablement or other unwanted bodily states which arise through other causes than disease cannot be called illness in Boorse's system. The biostatistical platform therefore is fundamental also for Boorse's notion of illness. (It may be observed that this interpretation of the concept of illness is not universal among English-speaking theorists. Consider, for instance, J.H. Marinker's [52] characterisation of the notions of disease, illness and sickness.)

The biostatistical theory is in many ways simple and elegant. It entails a scientific operationalisation of the concepts of health and disease. According to the theory we can in principle numerically determine the limits of the healthy and the diseased bodily functions. The theory is also very general. It applies not only to man but to all biological species. For every species, whether an animal or a plant, we can calculate statistically normal intervals and on the basis of these determine whether a particular individual is healthy or not.

In spite of these merits some very forceful criticisms have been raised against it. Let me here briefly summarise these:

1. The theory has not considered the great variation of normal values depending on environment, the individual's activities and her way of life in general.

2. The theory has not considered the mechanisms of compensation. A subnormal value of a particular function can be compensated for by a supernormal value of a neighbouring function with the same net result.

3. Many bodily or mental states which are intuitively considered to be diseases or signs of ill health, need not involve any statistically abnormal function. Given a certain very harsh environment, for instance a very hard climate, a certain reduction of function can in fact be statistically normal. (For a development of these arguments see my [71], pp. 23-33.)

4. Boorse is not himself completely consistent in giving a value-free account of health and disease. (This has been discussed in detail in Fulford [30] Ch.3.)

THE HOLISTIC CONCEPT OF HEALTH

As a consequence of this critical discussion there has emerged a completely different type of theory, which primarily aims at giving a positive characterisation of health without leaning on a previous understanding of disease and illness.[32] This type of theory is certainly not essentially new. Galen, for instance, said that health is a state in which we neither suffer evil nor are prevented from performing the functions of daily life. In a similar vein the American sociologist Talcott Parsons says: "Health may be defined as the state of optimum capacity of an individual for the effective performance of the roles and tasks for which he has been socialized".[33]

In most such holistic characterisations of health two kinds of phenomena are mentioned: first the subjective phenomenon of a certain kind of feeling, of ease or wellbeing in the case of health, and of pain or suffering in the case of illness; second the phenomenon of ability or disability, the former an indication of health, the latter of illness.

These two kinds of phenomena are in many ways interconnected. There is first an empirical, causal, connection. A feeling of ease or wellbeing contributes causally to the ability of its bearer. A feeling of pain or suffering may directly cause some degree of disability. Conversely, a subject's perception of her ability or disability greatly influences her emotional state.

Some would argue that the relation between the two kinds of phenomena is even stronger, i.e. that there are conceptual links between, on the one hand, a feeling of wellbeing and ability and, on the other hand, suffering and disability. According to this idea, being in great pain, for instance, partly means that one is disabled. Some degree of disability is here a necessary condition for the presence of pain, so that if a person's ability is not affected, she can be said not to be in great pain.

The assumption of a conceptual relation between pain and disability will be accepted in the present analysis: a person cannot experience great pain or suffering without evincing some degree of disability. But a person may have a disability, and even be generally disabled, without experiencing pain or suffering. There are some paradigm

32 It should be emphasised that this book deals exclusively with the notion of *human* health. For a characterisation of health in the context of non-human animals and plants, see the discussion in [71], pp. 139-143.

33 The quotation from Galen is taken from O. Temkin [101], p. 637. The quotation from Parsons comes from his [76], p. 117.

cases of illness where pain and suffering are absent. One obvious case is that of coma. Another is present in certain mental disabilities and illnesses. When a patient cannot reflect on her own situation, then her disabilities need not have suffering as a consequence. In short, wherever there is great pain or suffering there is disability, but the converse is not true.

These preliminary observations indicate that the concept of disability has a much more central place in the characterisation of illness or non-health than the corresponding concepts of pain and suffering. If only one of these characteristics is essential to the notion of illness, then disability is the primary candidate. This is my main reason for founding the subsequent analysis on the concepts of ability and disability.[34]

THE ABILITY OF HEALTH

What should a healthy person be able to do? What disabilities are such that a person considers herself in some respect disabled? What disabilities are such that she seeks medical care? And what disabilities are such that the health care system should deliver health care to her?

(These questions are clearly not identical. The latter question in particular has a political overtone. The answer to it depends not only on conceptual analysis but equally much on policy based on economic decisions concerning medical priorities. I shall not enter this discussion in this book.)

In approaching these problems I shall first somewhat alter my manner of speaking. Instead of talking about a set of actions that an agent must be able to perform, I shall assume that there is a set of *goals* which the healthy person must be able achieve. This does not involve a radical change in philosophy. It is merely a simplification of the mode of speech. By concentrating on (ultimate) goals I can avoid giving a long enumeration of specific actions. Moreover, I am not forced to make difficult decisions about the level at which action concepts should be specified.

It is plausible to believe that whatever the adequate answer to the question of the nature of health should be, it will be an answer on an abstract level, which can be summarised in terms of certain general goals. The question to be put should then rather be formulated in the following terms: what are the goals that a healthy person must be able to realise through her actions?

Let me, then, rehearse the logical relations between actions and goals. A first assumption is that the realisation of a goal requires the performance of at least one action. If the goal is general and abstract it normally requires a sequence of actions, i.e. a project. The relation between an action and its goal may be internal or external. The difference is easily seen given my analysis of actions above. We say that the relation is external if the goal is a causal or conventional *consequence* of the action. The relation is internal if it follows logically from the fact that the action has been performed, that the

34 In making this choice of conceptual platform I do not deny the importance of concepts such as pain and suffering for the understanding of the phenomenology of, what I call, "subjective illness". In my [69], pp. 103-112, I have dealt extensively with these issues.

goal has been reached. For example, the relation between working hard and passing an exam is external. Passing an exam is a causal consequence of working hard. The relation between killing Smith and the death of Smith, however, is internal. If Smith has been killed it follows logically that Smith is dead.

We see, then, that the way human action is conceptualised is crucial for the nature of the relation. One and the same sequence of episodes can be looked upon as either an action causing a goal (stabbing Smith *causing* the death of Smith) or an action entailing a goal (killing Smith *entailing* the death of Smith).

It is important to be conscious of these features for the following reason. The reaching of a goal can be conceived of as, simply, the performing of an action. However abstract a goal we choose to consider, we can always construct an action-concept entailing it.

It is now easily seen that the goal-mode of discourse is just a variant of the action-mode of discourse. Many of the conclusions drawn from the analysis of actions can be transferred to the analysis of goal-attainment.

When goals are far-reaching or abstract the process of attaining them is normally divided into a sequence of actions (a project). The endstate of each member of the project can be viewed as a subgoal. In the extreme case, the attainment of a goal requires the realisation of a very specific set of subgoals in a particular order. As was observed in the case of projects, however, most goals can be realised by a great number of alternative routes. Different circumstances may constitute opportunities for different actions in realising the same goal. In circumstance *C1* action *F1* may be necessary for realising goal *G*. In *C2* actions *F2* + *F1* may be necessary for *G*. In *C3*, finally, *F3* may be both necessary and sufficient for *G*.

But also one and the same circumstance may provide an opportunity for distinct actions leading or contributing to *G*. In such a case neither of them is necessary for *G*. Normally one can get food in many different ways. One may go out in the woods and pick berries, buy food in a shop, or go to a restaurant. Many persons have the opportunity of performing these different actions at any one time.

Observe that the number of alternatives is dependent upon the level of abstraction in our specification of actions. When the level of specification is very detailed the number of alternatives will be very great. Think for example of all the different routes one can take in walking to the shop in order to get food.

A further important question is the following: Is the ability of health a *first- or second-order ability*? I shall argue that the most plausible candidate is the latter. To be healthy, I shall suggest, is to have a second-order ability to realise one's vital goals. Let me defend this idea.

Compare the following two persons. John is a highly trained athlete. He has specialised in high jump and can normally jump 2 metres. One particular day, however, his leg aches and he can only jump 1.50. Peter is a man who has never trained high jump and when he tries the same day as John is injured he jumps 1.50. We assume, however, that Peter has the same biological capabilities as John. If Peter were to train he would also be able to jump 2 metres.

In this case John's and Peter's first-order abilities are identical at a certain time *t*. It seems, however, very implausible to say that both have the same degree of health. Intuitively it is John who is unhealthy and Peter may have perfect health.

Consider a further example. An uneducated young man leaves a developing country and moves to Sweden. In his native country he had his own farm, which he cultivated well enough to sustain himself and his family. When he enters Sweden, say as a political refugee, he is no longer able to lead such a life. Whereas in his home country he lived relatively well, in Sweden he is disabled. But would we say that this man is healthy in his native country, and becomes ill upon entering Sweden? No, it seems plausible to say that as long as he has the second-order ability, i.e. can successfully complete a language course and other necessary courses, then he may be completely healthy, also from the Swedish perspective.

In general, then, such disability as is solely due to lack of training is not an indication of illness. There is reason to speak of illness only if the act of training has in turn been prevented by internal factors, in which case there is a second-order disability.

But what about the typical case of illness which is due to an organic disease? Consider the following. An agent has a first-order ability to perform his professional activities. Then he becomes ill, and as a result loses his first-order ability. But would it be true to say that he no longer has the second-order ability to do his work?

It is easy to be misled here and identify two pairs of concepts which should be held distinct; one pair is first- and second-order ability, and the other is power to execute a basic competence and having a basic competence. We normally ascribe a basic competence to someone when he knows how to do something. According to our previous definitions this need not at all be true about second-order ability. The immigrant to Sweden has not previously learnt anything about Sweden and in this sense does not have the basic competence requisite for making his living in Sweden. He may, however, have a second-order ability with regard to the same action.

Basic competence is compatible with both second-order and first-order ability. But for our purposes it is important to see that a person who has a basic competence vis-à-vis a certain action *F* need not even have a second-order ability with regard to *F*. Consider a case of long-term illness. A professional footballer has broken both legs. It is obvious that during the ensuing period he does not have the first-order ability to play football. Still, we would say that he has, throughout this period, the basic competence to play football. He knows how to play football. But does this footballer have the second-order ability to play football while he lies in bed? No, for having the second-order ability to do *F* means having the first-order ability to follow a programme which leads to a first-order ability to do *F*. But the footballer, who is confined to bed, is clearly not in a position to follow such a programme; and so we may say of him that he is ill.

The same reasoning may be applied to all paradigm cases of illness due to disease or impairment. During an acute phase of illness, however short it may be, an agent has lost both the first- and the second-order ability to perform the action with respect to which she is disabled.

Is this intuition about the relation between health and second-order ability wholly tenable? Consider the typical case of *rehabilitation*. A person has been ill but has finally recovered. She is quite weak and still cannot do much. She thus starts a programme of rehabilitation. After this period of rehabilitation she is fit to do all that she could do before, both workwise and in her private life.

How should this process of rehabilitation be described? The most accurate way of describing it seems to be that she has gone from a state of second-order ability to a state of first-order ability. But given our conceptual machinery she has not then improved her health. But does not the process of rehabilitation, according to our intuitions, entail an inprovement in health?

I think there is a solution to this puzzle within my theory. The training involved in the rehabilitative process does not just entail a transition from second-order to first-order ability. It normally leads to a considerable improvement also in second-order ability. By exercising one's muscles in order to be able to run, one simultaneously improves one's capability for other exercises where these muscles are involved. It is the latter kind of change that strictly speaking constitutes the improvement in health, according to my analysis. By training a weak person in a rehabilitation programme one not only teaches her to do new things, one also teaches her to *learn to* do new things.

CHAPTER NINE

ON THE CONCEPTS OF VITAL GOAL AND HAPPINESS

I shall now confront the most crucial problem in the philosophy of health. What should a healthy person be able to do? What are the *vital* goals of man? And is there just one set of vital goals?

The latter question is a variant of the question whether there is only one concept of health, which I shall address in a special chapter. For the time being, however, I shall assume the possibility of finding an abstract and universally applicable characterisation of the set of human vital goals, which could serve as the platform for a definition of health. I shall first test this assumption by considering two basic candidates for such goals which have been suggested in the philosophical literature on health. The first suggestion is that the vital goals of a person should be equated with a person's fundamental *needs*. The other is that the vital goals should simply be equated with the subject's *wants* at a particular time.

I have elsewhere discussed both these suggestions in great detail and claimed that they are at best unhelpful and at worst completely inadequate.[35] My conclusions can be very briefly summarised thus:

On needs. "Need" is basically a relational term, more precisely a four-place predicate, where the terms are a subject, a situation, an object and a goal. For an illustration, consider the case of John who is placed in a situation where he needs a hammer in order to repair his house. More formally: A needs y in S in order to attain G. An alternative formulation, where "need" is a noun, is: There is a need of *A's* for y in S to attain G.

It is very uninformative, if not nonsensical, to say simply that John needs a hammer. Such a locution has to be expanded and clarified by answering the following questions. In relation to what goals have we identified this need? How are the goals determined? What is the situational background in relation to which the goals are identified?

This analysis is also relevant to the discourse where one talks of *fundamental human needs*. To what goal is a fundamental need related? In the perhaps best-known account of fundamental human needs, viz. Abraham Maslow's [54], one can (with some difficulty) trace the following answer to the first of the mentioned questions:

> The fulfilment of the basic needs is a necessary condition for the survival of the individual (or the species) or for the *health* of the individual.[36]

Only the first part of this disjunction can be used for the clarification of vital goals in the analysis of the notion of health. If the latter part of the disjunction is used we end

35 My previous discussion about the concepts of need and health is to be found in [71], pp. 57-65, and [69], pp. 137-145.

36 See A. Maslow [55], p. 21, for an elaboration of the idea that there is a connection between the concepts of need and health.

up in a circle: *A* is healthy means *A* has the ability to realise all those goals which are necessary for her health. As a result (using the system of Maslow's and a number of similar systems), fundamental needs can be related to survival only. This is, however, much too weak for our purpose. The healthy person must be able to do more things than just support her own survival.

On the subject's own goals. Could the vital goals of a person be identical with the goals that she actually sets for herself, i.e. what she intends or wants to realise? This simple suggestion has some initial plausibility since what the subject herself wants or intends to do is per definition of importance to her. (This is also an aspect on which I shall in the end base my own considerations.)

The suggestion can however not do unqualified, mainly for the following reasons. (i) The subject may set unreasonably high goals. Do we mean that all people with ambitions that are too high are ill? This does not seem to be a plausible conclusion. (ii) She may set unreasonably low goals. Does a person with very low ambition and very few goals — which she can easily achieve — automatically have a high degree of health? Can adequacy of resources for one's intended or wanted goals be a sufficient indicator of health in this person's case? This seems counterintuitive. (iii) She may set conflicting and counterproductive goals. People sometimes form intentions which are contrary to more highly valued preferences of theirs or which are even damaging to themselves. When a man decides to consume a great amount of alcohol, it is likely that his decision is in conflict with some other important aim of his, for instance that of performing his professional role properly. It seems counterintuitive to say that all kinds of irrational and counterproductive goals, just because they are intended, should have the status of vital goals.

VITAL GOALS AS GOALS RELATED TO LONG-TERM HAPPINESS.

My own solution to the characterisation problem rests on the notion of happiness. I shall say that a person's vital goals are *the states of affairs which are such that they are necessary and jointly sufficient for the subject's minimal long-term happiness*.

This proposal then presupposes a substantial analysis of happiness. Let me summarise the basic elements in my own analysis which has been inspired by E. Telfer [99], J. Griffin [35] and W. Tatarkiewiçz [97]. First an intuitive definition:

> *A* is completely happy, if and only if *A* wants everything in the world to be just as she finds it to be.

Happiness, as I understand it, is conceptually connected to the wants and goals of human beings. One is happy about the fact that one's wishes and goals are realised or are becoming realised. One can be happy about an academic achievement or an achievement in sports; or one can be happy about some external event which contributes to the realisation of some wish that one has. If one's life as a whole is characterised by the fact that ones's most important goals are fulfilled or are in the process of being fulfilled, then this life is with great probability a life in happiness or harmony. It is important to emphasise that the goals talked about here need neither be conscious to the individual nor be the result of a personal achievement or even constitute a change. It belongs to our most important goals to maintain the *status quo,* to keep our nearest and dearest with us, to keep our jobs and in general maintain our most fundamental conditions in life.

A more formal and precise way of expressing my basic idea, then, is the following:

A is completely happy at *t*, if and only if

(i) *A* wants at *t* that (*x1...xn*) shall be the case at *t*,
(ii) (*x1...xn*) constitutes the totality of *A*'s wants at *t*,
(iii) *A* is is convinced at *t* that (*x1...xn*) is the case.

According to this definition happiness could be said to be an equilibrium between the subject's wants and the world as she finds it to be. In most instances there is some disequilibrium between a person's wants and how she sees the world. This disequilibrium can be greater or lesser. Thus happiness must be a dimension extending from complete happiness to complete unhappiness. The latter is the state where nothing in the world is as the subject wants it to be.

The following important conclusions can be drawn from this conception of happiness.

(i) *Happiness is a partly cognitive concept.* This follows both from the conceptual connection between happiness and want and from the conceptual connection between happiness and perception. In order to be able to want something one must at least be a minimally intelligent being. If *A* wants to get a car, *A* must believe (or at least envisage) that there are such things as cars and that they are available. Wanting something and perceiving something entails some cognitive state, knowledge, conviction, belief or at least imagination that certain things are the case.

(ii) *Happiness is distinct from pleasure.* If happiness is a partly cognitive concept, pleasure is not. "Pleasure" is the term denoting a set of positive sensations, for instance the ones of tasting and smelling but also the ones concomitant with physical exercise. But since many of us want to have the sensations of pleasure we are normally also happy *about* our feeling pleasure. Thus pleasure is typically the object of happiness. Happiness is a second-order mental state that can take pleasure and other positive sensations, moods and emotions as objects.

The distinction between happiness and pleasure makes it easier to understand that there are cases when one can be unhappy about one's pleasure and happy about one's pain and suffering. To illustrate the first case, consider the man who dislikes the pleasure he gets from getting drunk the night before he is intending to take an exam. The pleasure is in this case a predictor of a future unwanted state which is going to have further negative consequences. Likewise a person in pain might realise that the pain is a necessary prerequisite for the attainment of an important desired end. A typical case of this is the medical situation where the patient understands that she must undergo a painful operation in order to be restored to health.

(iii) *Happiness as an emotion. A dispositional analysis of emotions.* Given my previous classification of feelings, happiness must fall into the category of emotions. In contradistinction to pleasure, happiness has *objects*, the significant feature of the emotions. The formal object of the emotion of happiness can roughly be described as "a state of affairs believed to be the case by *A* and believed by *A* to constitute the realisation of a want of hers."

The occurrence of happiness does not, however, entail that there is a feeling in the sense of a continuous experience during the whole period when it is true to ascribe happiness to the subject. Like many other emotions, happiness is consistent with the non-occurrence of any feeling whatsoever. Consider the extreme case of a sleeping person. It may be true to say of *A* that she is happy even when she is asleep.

On the other hand there is some analytical connection between happiness and feelings of happiness. Happiness is a disposition to have a a certain set of experiences. And normally the happy person also feels happiness during some parts of the period when she is happy.

On degrees of happiness

I have said that one can be either more or less happy with life. According to the want-equilibrium concept one can be completely happy with life, meaning that one wants all conditions which are affecting one's life to be exactly as they are. One is less happy if some conditions are not as one wants them to be, and one is completely unhappy if nothing is as one wants it to be.

How should this dimension of happiness be assessed and measured? Can the procedure be the following simple arithmetical one?

$$\frac{\text{Number of satisfied wants}}{\text{number of wants}}$$

I shall argue that this is a completely unsatisfactory idea. The reasons are two:

(i) It is doubtful whether wants can be counted in such a strict way that the above mathematics is reasonable.[37]

(ii) Wants have different *weights*. One want may be much more important than another want. *A*'s want to have children and her want to get a decent job have a higher priority

37 In [69], pp. 55-61, I discuss and cast doubt on the idea of counting wants.

than her want to have a good meal and her want to spend a holiday in Scotland. As a result *A* becomes happier if the former are realised than if the latter are.

I shall here explore the idea of *higher priority*. What does it mean to say that *A*'s want for *X* has a higher priority than *A*'s want for *Y*? It cannot simply mean that the satisfaction of the want for *X* gives *A* more happiness than the satisfaction of the want for *Y*. Then we would be moving in a complete circle. My explication is instead the following: *A*'s want for *X* at *t* has a higher priority than *A*'s want for *Y* at *t*, if and only if *A*, in a situation of choice at *t*, where *A* can choose either *X* or *Y* but not both, would choose *X*, unless prevented by external or internal force. (Observe here that *X* and *Y* need not be actions actually performable by *A*. The situation of choice must have the character of an intellectual experiment where the items to choose from can be states of affairs of any kind.)

With this formulation I can keep the connection to my basic analysis of the concept of happiness. I said that *A* is happy if, and only if, *A* wants the conditions in her life to be just as she finds them to be. I can now say that *A* is happier about a situation which contains *X* than about another situation which contains *Y*, and explain it by saying that *A* prefers the situation with *X* to the situation with *Y*.

Does this analysis allow for any comparisons between people? Can one ever with certainty say that *A* is happier than *B*? This can be done only given certain presuppositions which can be clarified in terms of our analysis.

(i) Assume that *A* and *B* have exactly the same profile of wants; they have exactly the same wants and also the same order of priority among the wants. Thus we know that both of them prefer *X* to *Y* (where *X* and *Y* are total situations). Assume then that *A* is in *X* and *B* is in *Y*. From here it follows that *A* is happier than *B*.

(ii) The requirements can be slightly relaxed. *A* and *B* may have different profiles, where, however, both profiles entail that *X* is to be preferred to *Y*. Then if *A* is in *X* and *B* in *Y* (where *X* and *Y* are total situations), *A* must be happier than *B*.

If this entailment does not hold or if *X* and *Y* are not total situations so that *A*'s and *B*'s lives contain further elements, which are not taken into account, then there is no comparison to be made on the basis of our analysis so far.

Thus in all comparisons — with the exception of a case to be discussed in the next section — between states of happiness both within one and the same person and between persons we must check that the profiles of wants are identical at least concerning the set of total situations under debate. If such conditions for comparison are not fulfilled I shall say that the different states of happiness are *incommensurable*.

THE NOTION OF ACCEPTABILITY AND MINIMAL HAPPINESS

There is, however, one important situation which is not uncommon and which is an exception to this incommensurability. This is the case when a person *A* finds her situation in life *unacceptable*, whereas *B* finds hers *acceptable*. In this case *B* is clearly happier than *A*. I hereby indicate the important borderline on the scale of happiness between happiness and unhappiness. I assume, then, that there is a level on the scale of happiness which marks a transition from happiness to unhappiness. Under this level the situation is so far from satisfactory that it is not acceptable; the person is unhappy. Just on the level or just above it the situation is acceptable; she is minimally happy.

In order for a person to be minimally happy in this sense a certain set of her wants have to be satisfied. I shall call this set the set of wants which have a *high priority* (in an absolute sense of the term).

ON THE POSSIBLE DEVELOPMENT OF COMPLETE HAPPINESS

Can the idea of happiness as an equilibrium be challenged? Consider the following problem. So far I have accepted the idea that a person can be completely happy, viz. in the case where she has all her wants satisfied. But what kind of completeness is this? Does this preclude a person from becoming a happier person in the future? And how could this change for the better be characterised within my theoretical framework?

In a previous work [69] I have introduced a further dimension of happiness in order to allow for this possibility. I called this dimension *the dimension of richness*. To illustrate this case I chose the following example.

A boy has lived all his life in simple and unpretentious circumstances in the Highlands of Scotland. He has been entirely content with his lot; he gets along well with his family, he appreciates the wild countryside and he enjoys the sometimes hard struggle for life up there. Thus he has been completely happy in the want equilibrium sense analysed above. One day he and his family are visited by a tourist who happens to be a famous musician. The tourist is attracted by the place and settles there permanently. Partly to earn his living he starts teaching the young boy to play the violin. He then discovers that the boy has a remarkable talent for music and that he very soon develops a proficiency in playing. This completely changes the boy's life. A whole new world has been opened to him. He enjoys playing and he enjoys listening to music. To put it in my technical terms, he has acquired a number of wants which he did not have before, and he is in the process of satisfying them.

The boy was, as I said above, completely happy with life before the musician arrived. But how should I then express the positive change that has now happened to him? Let me describe the situation formally. *A* wants at *t* the states *x*, *y* and *z* to be the case. *x* , *y* and z happen to be realised at *t* and *A* is completely happy. At *t1* *A* gets to know about the existence of a new state of affairs *e* and wants to have that. He still wants *x*, *y* and *z*. Assume now that *x*, *y*, *z* and *e* are realised at *t1*. And assume also that there is nothing further realised that *A* does not want to have. Is *A* as a result necessarily happier at *t1* than at *t*?

According to my present analysis we cannot in general draw this conclusion. We can, for instance, envisage the following case. It may be true that *A* at *t1* still wants *x*, *y* and *z*, which she also wanted before. But — presumably as a result of his strong want for *e* — *x*, *y* and *z* may have declined considerably in importance for him. Thus, it is not so clear any longer that *A* is happier in the later situation than in the earlier one.[38]

This difficulty does not entail that we cannot, in some situations, claim that a person is happier at time *t1* than at time *t*, although she was at *t* completely happy. We must then, however, as before rely on the notion of preference. We must furthermore presuppose that *A*'s preference structure is identical at *t* and *t1*. The conditions can be formalised in the following way:

Assuming that *A* is completely happy at *t*, *A* can still be said to be happier at *t1* than at *t*, if and only if

(i) *A* prefers at *t1* her total situation at *t1* to her total situation at *t*,
(ii) *A* would at *t* prefer her total situation at *t1* to her total situation at *t*, had she at *t* been adequately informed about what the total situation at *t1* would be like.

Thus the notion of complete happiness in the equilibrium sense is, as I now see it, a relative one. The completeness is relative to the set of wants that a person has at a particular moment. It does not take into account the person's potential wants. From a conceptual point of view there is no *absolutely* complete happiness. The only limit to happiness, then, is an empirical one. A person cannot have more than a limited set of wants and she cannot realise more than a limited set of wants.

Objections to the Suggested Analysis of Vital Goals and Happiness

My characterisation of the two crucial notions of vital goal and happiness may still warrant some justification. I have elsewhere attempted to do this in a more thorough way [71], pp.81-104 and [69], pp. 55-80, but I shall here rehearse three major points.

(i) On the difference between wanted goals and vital goals

I earlier dismissed the idea that a person's set of wanted goals could be identified with her set of vital goals. Instead I suggest that vital goals are related to the subject's long-term happiness. Happiness, in its turn, however, is defined in terms of the realisation of the subject's wants. Am I then simply moving in circle?

No, the answer is that the first suggestion for the characterisation of vital goals is unacceptable because it does not at all discriminate among wants. It treats on a par trivial wants, counterproductive wants, destructive wants and essential wants. The concept of happiness-in-the-long-run identifies instead those states of affairs — indeed deeply wanted by the subject — which last for a long time or contribute to a development over time, which in its turn is still wanted by the subject. Trivial, counterproductive or destructive states of affairs — even if they have at one time been wanted — are therefore excluded from the set of vital goals.

38 For a discussion about the richness dimension of happiness, see P. Sandøe and K. Kappel [86] and my own reply in [69], pp. 71-73.

(ii) On the reference of wants to future states of affairs

At a particular moment in time a person *A* has a variety of wants. Some of these refer to the present time, others refer to some time in the future. *A* may *now* want *X* to be the case *now* and *A* may *now* want *Y* to be the case in five years time. It is crucially important for *A*'s happiness now that her wants which refer to the present time are satisfied. *A* may, for instance, want her children to be now in good health and do well at school. *A* is happy (or at least it contributes to *A*'s happiness) if *A* sees that this is true of her children now.

For obvious reasons those of *A*'s wants which relate to the distant future need not (or sometimes should not) be satisfied now. *A* may want to become the Prime Minister of Sweden in five year's time. *A* does not want this to be realised now. Thus it will not, given the premises, contribute to *A*'s happiness should she by any chance get this position now. But what is then the relation between happiness and such unsatisfied wants as refer to some future time? Should the latter be excluded from the discussion?

The answer is that there is an important indirect way in which future-related wants also affect present happiness. If the present conditions are such that they do not favour the realisation of *A*'s future-related wants, or if they even prevent this, then *A*'s happiness will be affected negatively. *A* is happy now only if *A* believes that the prospects of the realisation of her future-related wants are good. Those prospects can be viewed as conditions existing now, which *A* wants to be the case.

(iii) On the plausibility of a want-satisfaction theory of happiness

It could be maintained that the theory presented in this study — purporting to be a general theory of happiness — is only a theory of simple satisfaction. Happiness is here in fact equated with the satisfaction of wants. This, it may be said, is a very crude and simple theory. Happiness, and human well-being in general, is something much more complicated and profound.

Consider a devout Christian who argues along the following lines: The happiest life is not a life of want-satisfaction; it is the life which is in the vicinity of God, in the Mass, in Holy Communion or in prayer. Analogously a scientist or an artist can claim that the really happy life is the life of science or the life of artistic creativity. Such lives have qualities which make them more than worth living, in comparison with the bureaucratic and grey lives that most of us live, and they are not characterised by simple want satisfaction.

There is a first preliminary answer to this. The religious, scientific or artistic lives are happy lives for those who have the interest in and inclination for such lives. To a devout Christian the religious life is splendid, to an intellectual the scientific life may be splendid etc. But having introduced the notion of interest we have in fact reintroduced the notion of want. It is, for instance, a fundamental want of the religious person to live in the vicinity of God. Everything else has a much lower priority.

However, the devout Christian may not just think that people who are interested in religious life will become happier by living religiously; she may think that everybody, interested or not, will become happier. The religious case is here stronger than the artistic and scientific cases. It is a fundamental hypothesis in most religions that it is possible for everybody to live a religious life to some extent.

I disregard here whether the religious claim is true or not. The general idea is plausible enough. There may very well be a set of activities such that (almost) everybody, whatever her gifts or talents, would find indulging in them highly enjoyable or — here we may want a stronger expression — highly rewarding or engrossing. Only a few people know about this fact, only a few people actually want to take part in these activities.

I think that we can give an account of this case also in my own framework. Let me first distinguish between the happiness aspect and the pleasure aspect of the engrossment talked about here. We may grant that the religious person has almost sensation-like feelings of extreme pleasure. But pleasure, I have argued, is not a part of happiness. We need independent arguments for saying that the religious person is also happier. But this is, indeed, also possible.

The religious person has a strongly positive attitude to life. She sees the meaning of life quite clearly and she has a positive attitude to her peers and mankind as a whole. This attitude involves frequent cognitive acts, frequent reflections on how well God has organised the world for everybody. In fact, the person may think that God has organised the world in a way which is such that one could not want it better.

In the case we are imagining then, the religious person thinks not only that God has satisfied all her (legitimate) wants; she thinks more profoundly: God has satisfied all conceivable (legitimate) wants; the world could not be better. This, I would say, is certainly an ultimate kind of happiness. But I maintain that it is still expressible in the terms proposed in this book.

On the Relation Between Health and Happiness

Let us now recall the place of happiness in the theoretical framework of health and vital goals.

The notion of happiness was introduced in order to characterise the notion of a vital goal. The latter in turn is a central component in my notion of health. I can then reformulate my definition of health in the following way.

> *A* is completely healthy, if and only if
> *A* is in a mental and bodily state which is such that
> *A* has a second-order ability, given accepted circumstances, to realise the states of affairs which are necessary and together sufficient for *A*'s minimal happiness in the long run.

This can look like a very strong connection between the two concepts of health and happiness. However, it is important to observe exactly what it entails. This relation between the concepts allows, for instance, the following. A person can be completely healthy and at the same time be very unhappy. The circumstances under which the healthy person lives can be extremely unfavourable. The healthy person who lives in the circumstances of war or who has lost a near relative is rarely a happy person. Moreover, the healthy person can abstain from using her ability to reach the vital goals.

But the characterisation also allows that a person with a very low degree of health can be very happy. A person with a low degree of health and a small ability can be compensated for her disability, so that the vital goals can still be reached. This is in fact what the doctor or nurse tries to achieve when providing a patient or client with the things needed. The compensation can, for instance, entail that the patient/client receives technical devices which will enable her to do what she finds important.

APPENDIX: SEN ON FUNCTIONINGS AND CAPABILITIES. A RIVAL THEORY OF WELL-BEING

Introduction

In this section I wish to relate my theories of health and happiness to the well-known theory of well-being which Amartya Sen has proposed and developed in a great number of publications. On the surface there are interesting similarities between Sen's emphasis on so-called capabilities and functionings in his characterisation of welfare and my own choice of abilities in characterising health. What then needs to be asked is whether my notion of health is closely similar to Sen's notion of well-being and, what conclusions, if so be the case, can be drawn concerning the reasonableness of the two theories. My subsequent analysis will show that the similarity between our two theories is quite superficial. Sen's version of well-being is quite far from my view of health. For one thing, Sen's notion of capability is much broader than my notion of ability. In order to show this I can use some elements from the conceptual apparatus introduced earlier in this book.

On the topic of well-being our respective proposals are quite different. Sen suggests a "midfare" theory, i.e. a theory which locates well-being between, on the one hand, the set of external resources, such as money, other economic assets, or social status, and the mental results of exploiting such resources, such as pleasure and happiness. My own view, as has been shown earlier, entails locating well-being at the latter end.

But let me first give a brief outline of Sen's theory of well-being insofar as it relates to the basic central concepts. I shall not attempt to follow him, however, in his more technical applications which are mainly geared to the theory of economics. In my presentation I shall basically refer to one of his more recent writings, viz. *Inequality Reexamined* [92]. Reference will, however, also be made to other of his writings, in particular *Commodities and Capabilities* [91].

It is important to place Sen's theory of well-being in its context in the modern debate about equality and welfare. Sen reacts to two tendencies in modern (and indeed also ancient) theorising on welfare. One tendency is the classic utilitarian, which identifies welfare with certain end-states of actions, normally within a person's mind. According to the utilitarian theories, in their most famous hedonistic versions, the welfare or well-being of a person is identical with the amount of pleasure or happiness that this person possesses at a certain time or during a certain period. Thus pleasure or happiness is the thing to be measured in evaluations of equality of welfare. A theory of this kind has, according to Sen, the serious disadvantage that it can equal the well-being of a content beggar with the well-being of a billionaire. Such a result is absurd, he contends, if we wish to build a theory of welfare that could be of any use in, for instance, political theory.

Another tendency goes in the completely opposite direction. Theorists in this quarter attempt to identify people's welfare with the amount of valuable goods or resources that these people have. A famous theory in this category is that of John Rawls [83], who identifies a certain set of *primary goods,* the possession of which determines the owner's state of goodness. (Rawls does not use the terms "welfare" and "well-being".) Among these goods are certain social ones, such as rights and liberties, whilst others are natural goods, including health, wealth and income. It is these goods that should be distributed in a specific way in order to create a just society.

Such a theory, says Sen, is also inadequate, since it does not take account of people's varying abilities to take advantage of goods. A person who cannot read cannot use all the brilliant books that she may possess. A person who is completely uneducated cannot take advantage of all the moden technology that the modern Western society provides. Thus the goods in themselves cannot constitute welfare, according to Sen. The goods are only good if they can be used by some person. Thus the kernel of welfare lies in the usage of goods, not in the goods themselves.

Sen's theory of functionings and capabilities

The well-being of a person, Sen says, can be seen in terms of the quality of the person's living. Living, in its turn, consists of a set of interrelated *functionings*, consisting of beings and doings. "The relevant functionings can vary from such elementary things as being adequately nourished, being in good health, avoiding escapable morbidity and premature mortality etc., to more complex achievements such as being happy, having self-respect, taking part in the life of the community, and so on" [92], p.39. An evaluation of a person's well-being must, according to Sen, take the form of an assessment of the constitutive elements of living, viz. the person's various functionings. Whether a person is well-nourished, in good health, etc. is intrinsically important for the wellness of that person's life, as Sen puts it.

Although functionings in a way define the life of the person and the evaluation of functionings therefore is an evaluation of the well-being of the person, Sen strongly wishes to add a further dimension, which is highly relevant for the person's well-being. This dimension is the *capability* dimension of the person.

A person's capabilities are related to her well-being for at least two reasons, says Sen:

1. The capabilities of achieving functionings (i.e. all the alternative combinations of functionings a person can choose to have) constitute the real freedom — the real opportunities — to have well-being.

> This "well-being freedom" may have direct relevance in ethical and political analysis. For example, in forming a view of the goodness of the social state, importance may be attached to the different freedoms that different people respectively enjoy to achieve well-being. Alternatively, without ta-

> king the route of incorporating well-being freedom in the "goodness" of the social state, it may simply be taken to be "right" that individuals should have substantial well-being freedom [92], p. 40.

The importance of capabilities in Sen's theory is visible also when he introduces the notion of an *evaluative space.* The evaluative space is defined by the set of objects which are to be evaluated in the assessment of a person's well-being. The evaluative space of utilitarians, he says, consists of utilities (normally certain states of mind). His own evaluative space consists of functionings and capabilities of functionings.

Thus, Sen seems here to advance an enlarged theory of well-being. According to this theory, not only functionings but also capabilities of functionings, i.e. people's freedom to achieve functionings, belong to the range of well-being. At least this seems to be the case when we describe well-being on the aggregated level of the state.[39]

Whatever Sen's final word on the issue of characterisation, capabilities become important in his theory of well-being simply because of their instrumental utility. A capability of becoming rich is important because it is a necessary condition for becoming rich, the prosperity itself being highly valued. But this is not the only relation between the two.

2. The second connection between well-being and capability takes the direct form of making achieved well-being itself depend on the capability of functioning. "Choosing may itself be a valuable part of living, and a life of genuine choice with serious options may be seen — for that reason — richer. In this view, at least some types of capabilities contribute *directly* to well-being, making one's life richer with the opportunity of reflective choice" [92], p.41. So when capability is understood as the actual process of choosing it contributes directly, according to Sen, to the well-being of the individual. In this case it seems to be clear, then, that characterising the capability in question *is* a part of the evaluation of the well-being of the person.

So much for the ontological categories which constitute the evaluative space of well-being. But what is the scale of evaluation to be used? What capabilities and functionings are valuable from a well-being point of view? In this respect Sen is surprisingly silent. As far as I can see he has in this essential respect abstained from providing a complete theory of well-being of his own. What we have are a few and indirect, and mostly negative, clues:

1. Sen discusses the relationship between *agency freedom in general* and *well-being freedom* in particular. He observes that not everything that a person wants and chooses to do is included in her well-being. A person can, for instance, choose to help the population of Rwanda. The state of the population of Rwanda cannot be seen as part of the well-being of this person. Thus general agency freedom is something broader than

---

39 Consider Sen's statement in [92], p 46: "The primary claim is that in evaluating well-being, the value-objects are the functionings and capabilities". But "that claim neither entails that all types of capabilities are equally valuable, nor indicates that any capability whatsoever — even if totally remote from that person's life — must have some value in assessing that person's well-being".

well-being freedom. However, says Sen, "a person's goals will, in the case of most "normal" people, include *inter alia* the pursuit of their own well-being. Indeed the overall balance of agency objectives might be seen, with some plausibility, as reflecting the weights the the person herself would attach to her own well-being *among* the things that she wishes to promote" [92], p.69. Although Sen here admits a strong relationship between agency freedom in general and well-being freedom, he points out a number of cases when the two can go in different directions. An increase in one's ability to promote goals that one has good reasons for promoting, such as sending food to the third world, may decrease one's ability and opportunity to spend time to enhance one's personal well-being. Sen also notices the interesting case where the expansion of one's freedom in itself makes life more difficult to live. It provides the person with a greater number of options. The person, however, finds herself disturbed by all these options. They become a burden and prevent the living of a good life.

2. In *Commodities and Capabilities* [91], Sen discusses the relationship between desiring and valuing. He distinguishes the following two propositions: (I) I value *x* and so I desire it, and (II) I desire *x* and so I value it. He then goes on:

> The approach that is being developed in this monograph is perfectly consistent with (I), and indeed frequently enough desire may well be taken (by observers) as *evidence* of value. But it is not in line with (II), which leans in a particular utilitarian direction. The difference is a foundational one, and relates to the question as to whether desiring is itself a valuation, or at least itself a source of value. This can be denied ... without rejecting the relevance of desire infomation in bringing out underlying — and often implicit — valuations [91], p. 32).

The negative conclusion here is clear enough. There seems however also to be something positive implied. Sen is in this example completely focused on the *subject's* desires and the *subject's* valuations. He seems here to presuppose that if a man *A* values *x,* which he possesses, then *x* is a part of *A*'s well-being. The *subject's* valuation of *x* is then one (but clearly not the only one; see Sen [91], p.33) criterion of *x*'s being part of the subject's well-being. If this is Sen's position it has some substance. It invites further questions and clarifications, though. We must first add that the valuation made by the subject cannot be any odd valuation. We have already noticed that *A* can morally valuate a certain way of acting, and proceed to act in accordance with this valuation (for instance, in the case of sending food to the third world), without this contributing to *A*'s personal well-being. The relevant valuation performed by the subject must then be *a valuation of the well-being kind.* This then begs the question. We want to know what well-being is.

An analysis of functionings and capabilities

Let me now attempt to analyse Sen's notions of functioning and capability in order to place my own theory of health in relation to it. Unfortunately there are no explicit definitions of the two concepts. Although Sen has written a lot about them both, partly in polemics with other writers, he chooses — no doubt intentionally — to keep the notions fairly open. He is still presenting a sketch; he seems to be waiting for a final closure of the concepts.

In my introduction I gave some examples of functionings (presented in Sen [92]). Let me here start off the analysis with a theoretical presentation in Sen [91]:

> In getting an idea of the well-being of the person, we clearly have to move on to "functionings", to wit, what the person succeeds in *doing* with the commodities and characteristics at his or her command. For example, we must take note that a disabled person may not be able to do many things an able-bodied individual can, with the same bundle of commodities. A functioning is an achievement of a person: what he or she manages to do or to be. It reflects as it were, a part of the "state" of that person. It has to be distinguished from the commodities which are used to achieve those functionings. For example, bicycling has to be distinguished from possessing a bike. It has to be distinguished also from the happiness generated by the functioning, for example, actually cycling around must not be identified with the pleasure obtained from that act. A functioning is thus different both from (1) having goods (and the corresponding characteristics), to which it is posterior, and (2) having utility (in the form of happiness resulting from that functioning), to which it is, in an important way, prior (pp.10-11).

Again the general idea is fairly clear. We must distinguish functionings from external goods, such as money, houses and bicycles. Moreover, functionings should be distinguished from those end-states of actions which other theories favour, such as happiness and desire-fulfilment.[40]

When one tries to get a sharper understanding of the ontology of functionings, difficulties arise. It seems to me that there is a tension in Sen's writings between a very wide (which I take to be the most plausible one) interpretation of functionings, including all kinds (or practically all kinds) of beings and doings of a subject, on the one hand, and a narrower interpretation, where there is a focus on the subject's *achievements*, i.e. a more action-oriented concept. The quotation just given represents the narrow interpretation: "A functioning is an achievement of a person: what he or she manages to do or to be". Sen's later work indicates something wider. He there explicitly talks about being in general. He exemplifies with being well-nourished and being in good health, both states which need not presuppose any specific action on the part of the subject.

In my reconstruction of Sen's concept of functioning here I shall choose the wider interpretation.[41] Functionings are both doings and beings. All kinds of actions, as I have characterised them in this book, are functionings. Moving a hand is a functioning, taking a walk is a functioning and so is signing a contract. But also the results of actions, such as having one's hand in a certain position, being in a new place and standing in a new legal relationship, are functionings. (However, not all results of *A*'s actions are functionings of *A*'s. Some results lie outside *A*'s body and mind and may cease to have a relation to *A*, for instance a ship built by *A* and sold to a company far away.) Other examples of functionings are beings which have no relation to any action whatsoever. Being tall is a functioning, being shy is a functioning and so is being old. But the matter does not stop here. Becoming taller, becoming shyer and becoming older, i.e. processes, are functionings. It is reasonable to assume that functionings include all kinds of epi-

---

40 A minor problem crops, however, up already here. In the later text [92], happiness is referred to as a special complex type of functioning. Thus happiness is not excluded from the set of functionings. It is however not awarded a privileged place among functionings.

41 This, indeed has support in a late writing of his [93].

sodes as I have characterised them in this book: states, processes and events, including agentive states, processes and events. So, as long as a subject *A* is involved in an episode, be it agentive or non-agentive, this episode is a functioning of *A*'s.

I think it is reasonable (and I think that it is ultimately his own contention) to include all episode-types in Sen's functionings. All episode-types can be relevant from a well-being point of view. A person's height and weight can certainly be important, a disease process can be highly important, so can a sudden event which deeply involves this person. The relevance of the subject's own actions for her well-being is self-evident.

These observations have important repercussions for the notion of capability. Capability is much more than ability in an action-theoretic sense. Sen has said so explicitly, having been attacked on this point.

> The crucial Greek word used in this context (by Aristotle among others), namely *dunamis,* can be translated as "capability of existing or acting", and presumably "existing" need not be the result of some vigorous "exercise of capability" [93], p.44.

So, what does "capability" stand for? It evidently covers "ability" in the sense specified in this book. But it certainly also covers "opportunity". Sen says in [ ], p.44: "A person's ability to achieve various valuable functionings may be greatly enhanced by public action and policy, and these expansions of capability are not unimportant for freedom for that reason." Thus, capability expands by external policy, for instance by giving people a wider and more suitable platform for action. And a platform for action is precisely what the term "opportunity" stands for.

But this is not all. The reference to Aristotle above indicates that "capability" also stands for such preconditions for functionings as have nothing to do with intentional action. External and internal conditions for becoming tall, for sleeping well, for having a healthy body, are all capabilities in Sen's broad sense. So, as far as I understand, everything covered by my term "practical possibility" for action, but also "practical possibility" for any other kind of episode pertaining to a person *A*, is a capability of *A*'s.

To put matters in a simplified way: A functioning is any property (apart from logical necessities) of a person and a capability is any precondition for having such a property. Regarding this we can make a set of interesting observations. Since all properties are at the same time preconditions for having a large set of other properties, all functionings are also capabilities of other functionings. Being well-nourished, which is a functioning, is a precondition for working well, which is another functioning. It is easy to see that all functionings are also capabilities with respect to some other functioning. However, not all capabilities are functionings. This is so since not all capabilities are properties of the person in question. The configuration of states external to the person are also capabilities. They are, however, per definition, not functionings of the person.

This circular dependence makes it sometimes difficult to differentiate between the "real" functionings and "real" capability states, if we should make such a differentiation. Every capability is a capability in virtue of the existence of a state of affairs which can be described in non-capability terms. If one is able to win an Olympic race or is able to write a book, this is so in virtue of a set of conditions. Some of these conditions are functionings.

Comparison between the capability theory of well-being and the ability theory of health

Let me sum up the essential elements of Sen's theory of well-being. Sen has in his theory indicated what he calls the evaluative space of well-being. This means that he has indicated what types of objects should be ranked and evaluated from the point of view of equality of well-being. These objects are functionings and capabilities. In some writings functionings seem to be the primary objects of evaluation. In others the two types are treated as equal.

My primary aim in this chapter has been to analyse the concepts of functioning and capability and look at their interrelations. I have found that — in spite of some "action-oriented" formulations, which have misled certain commentators — Sen's concepts are extremely open and wide. A functioning of a person, I have concluded, is any property of this person. A capability of a person is any precondition of a functioning of the same person. Moreover, all functionings are at the same time capabilities of other functionings.

These conclusions enable me now to compare Sen's theory of well-being (as far as the evaluative space is concerned) with my own theory of health.

1. The evaluative space of Sen's theory of well-being is much broader than the constituents of health in my theory. This is an expected and, indeed, highly reasonable result. Well-being must concern something much more than health.

2. Some of the constituents of Sen's theory of well-being are identical with the main constituents of health, according to my conception. A person's abilities to achieve vital goals are certainly important capabilities in Sen's theory.

This result only reflects Sen's general aim of identifying well-being on a person's "midfare" level. Sen has explicitly mentioned health among his functionings. Probably, however, he did not regard health so much from the capability perspective as I do. The difference between us here need not be substantially great. I attempt to make the relation between health and well-being (happiness) as clear as I can, indeed, by making the two notions analytically related. A person's health is her ability to realise vital goals. These vital goals are in their turn necessary conditions for the person's long-term happiness. Sen has so far, to my knowledge, left the relation between specific functionings and well-being quite open, since he has not provided us with any specific value theory.

Comparison between the capability theory of well-being and the want-fulfilment theory of happiness

The capability theory of well-being is in a much clearer sense a rival to the kind of want-fulfilment theory of happiness which I have been advocating earlier in this chapter. This is so partly because Sen has explicitly taken issue with the latter kind of theory. Let me here review Sen's analyses on this point and see what answers can be made to his criticisms.

Let me first observe that Sen distinguishes between happiness and desire-fulfilment. He seems, then, to have a hedonistic theory of happiness in mind, a theory where there is no clear distinction between pleasure and happiness. For the particular arguments raised by Sen in this context we might disregard Sen's distinction between happiness and desire-fulfilment. Sen says for instance the following:

> Both the views of utility have the twin characteristics of (1) being fully grounded on the mental attitude of the person, and (2) avoiding any direct reference to the person's own valuational exercise — the mental activity of valuing one kind of life rather than another. The former I shall call "physical-condition neglect" and the latter "valuation neglect" [91], pp. 20-21.

Sen goes on to give an example of the physical-condition neglect.

> A person who is ill-fed, undernourished, unsheltered and ill can still be high up in the scale of happiness or desire-fulfilment if he or she has learned to have "realistic" desires and to take pleasure in small mercies. The physical conditions of a person do not enter the view of well-being seen entirely in terms of happiness or desire-fulfilment, except insofar as they are *indirectly* covered by the mental attitudes of happiness or desire [91], p.21.

I find this statement to be very interesting and worthy of a thorough analysis. I think that it clearly reveals that the author is a well-being theorist who has a social science platform, and not one having an individualistic or psychological platform. To Sen it seems *self-evident* that the person who is undernourished, unsheltered and ill, has a low degree of well-being. A theory which were to allow such a person to have a high degree of well-being must automatically be wrong, Sen seems to say.

I, for one, do not find his example to constitute a conclusive argument. However, I still find it to be an important point. A happiness-theorist like myself must articulate my view in relation to an example like his. Let me here attempt to do this stepwise.

1. First, the undernourished, unsheltered and ill person is in most instances a very unhappy person. Most persons do not want to be undernourished, unsheltered or ill. A person in any of these circumstances is normally a suffering person, and suffering is almost always an unwanted state of affairs. Moreover, the fact that *A* suffers, in the context of poverty, normally functions as an important, if not necessary, criterion for the hypothesis that *A* is undernourished or ill. Thus, in this typical case, Sen and I can agree that the person lacks well-being.

2. Second, even if the undernourished, unsheltered and ill person is not at present suffering there is, if the situation is grave, a high probability that she will suffer within a short time. Our empirical knowledge tells us that certain degrees of undernourishment, lack of shelter and illness must, unless treated, sooner or later lead to pain or other kinds of suffering. Thus, on my view, the undernourished etc. person is seriously *endangered* with respect to her happiness; the person's well-being is threatened. There are very good reasons for taking action and helping this person. Hence, from a pragmatic point of view Sen and I would come to similar conclusions also in this case.

3. Third, let me then consider the case of an undernourished, unsheltered and ill person who does not suffer, or for other reasons does not want to change her predicament, and whose situation is such that she might avoid suffering in the future. What is Sen's position concerning this person? If I understand him correctly we must still consider this person as severely lacking in well-being. The functionings of nourishment, being sheltered and having health are so important in themselves (having an absolute value?) that whatever the person's expressed wishes or lack of wishes, it would be absurd to attribute a high degree of well-being to this person. In judging this case, if I understand Sen correctly, he and I part company. I can imagine circumstances when this person has a high degree of well-being. And I would, accordingly, find a theory of well-being which cannot account for this possibility to be a defective theory.

Consider some cases.

a. The conditions of undernourishment etc. are not so severe. The subject is a strong person and can easily compensate for these conditions. She is doing well at work and is in general flourishing as a person. To make my point more precise: This person, although undernourished according to a normal scale, lives the same kind of life, and enjoys life as much, as her comrades who have a richer and more varied nourishment. This seems to me to be a clear case where the absolutist approach becomes absurd.

b. The subject is a man living in a poor country which is led by a despotic regime. The external circumstances are poor in this country. Most of his fellow-countrymen are indeed suffering. The man and some of his comrades have however found a successful way of resigning to the external circumstances and have concentrated on developing some of their talents and interests which it is still possible to cultivate in this country. Hence they have, paradoxically enough, become quite happy. They have come to know much about themselves and they in fact, in retrospect, value the harsh circumstances which have made them grow as human beings.

One may counter here that Sen has certainly not had such examples in mind. If the persons involved grow as human beings, then there is a case of high well-being also according to Sen. But if Sen were to counter in such a way, then his theses about the undernourished, unsheltered and ill persons do not have the absolute character that his wording quoted above seems to indicate. Resignation may be a way to growth. And if Sen accepts this fact, and also accepts that growth is a positive functioning from a well-being point of view, then he must tell us the criteria by which to judge such a fact. It seems to me very difficult to see how he could ultimately find such criteria without introducing some final-end theory of value.

One must remember that much work in health care and social care entails helping people to become realistic in order to cope with life. And I believe that Sen considers health care and social care in general to be work for the well-being of people. It is often difficult to change people's circumstances, regarding economy, employment, housing and sometimes also regarding basic health. It is most of the time possible, however, to help people psychologically. One of the best means is to help develop their attitudes to and expectations from life. By doing so one may contribute to a radical improvement of their happiness. According to my conception of well-being this is tantamount to saying that one contributes to their well-being. But Sen, on the other hand, given his theory, seems to be forced to claim that their level of well-being cannot be raised simply by psychological means.

In order to make Sen's position more understandable I think one must take him as attacking a very special and quite unsophisticated theory of desire-fulfilment. Sen says in *Commodities and Capabilities* [91]:

> Considerations of "feasibility" and of "practical possibility" enter into what we dare to desire and what we are pained not to get. Our mental reactions to what we actually get and what we can sensibly expect to get may frequently involve compromises with a harsh reality. The destitute thrown into beggary, the vulnerable landless labourer precariously surviving at the edge of subsistence, the overworked domestic servant working round the clock, the subdued and subjugated housewife reconciled to her role and her fate, all tend to come to terms with their respective predicaments (p.21).

Sen seems to require that a desire is a mental state, which, if not verbally expressed or otherwise made explicit, is at least conscious to its bearer. The phrase "we dare to desire" indicates this. My own concept of want does not require consciousness. (Consider my dispositional analysis of wants in chapter 4.)

Sen is also forced to have quite a strong conceptual requirement with respect to desires to make his own examples at all plausible. Consider the labourer, precariously surviving at the end of subsistence, or the overworked domestic servant. Could one even describe the situations of these persons in this way, unless they suffer from their predicaments? And if they suffer, they would almost necessarily want some change, unless there are specific reasons to the contrary (although they may not know exactly what to want). It is possible that the bearer does not dare to *express* her wants, or that she does not dare to make them *conscious*. But it is hardly conceivable that she does not want any change at all.

Theories like my own are often attacked on the ground that they are supposed to have dangerous political implications. They may even be supposed to serve totalitarian regimes or in general unequal societies. I think this is a grave misunderstanding. To point out the psychological route to well-being is not the same as regularly recommending it. I certainly favour political solutions to poverty, inequality, repression and other sources of human unhappiness. The only thing I wish to say is that there exist other ways to enhance the well-being of people if such solutions are non-existent.

Consider now the second serious shortcoming of want-fulfilment theories, viz. valuation-neglect. Valuing is not the same thing as desiring, says Sen. The well-being of a person is dependent on the achievement of what she values, not of what she desires. I have above already commented on Sen's distinction between valuation and desire. I shall repeat some conclusions here.

I think I agree about one reason behind Sen's distinction. This is the observation that not any odd desire can guide us to long-term well-being. On the other hand I think Sen introduces a great problem with his distinction. A person can have desires and wants of quite different kinds and of quite different dignity. A man can want to scratch his nose, he can want to have a bottle of whisky, he can want to seduce his neighbour's wife, he can want to help his daughter at school and he can want to get a prestigious job. These wants differ with regard to the manner and degree in which their satisfaction contributes to *A*'s well-being. Some provide momentary pleasure, others may contribute to lasting satisfaction. The realisation of some of the wants may in the long run indeed be detrimental. The drinking of a bottle of whisky may ruin his next day and the seduction of the neighbour's wife may ruin his life. All this is very clear. I have earlier in this chapter tried to sort out some complications in this area. In short, the satisfaction of all wants contributes to some extent to the bearer's happiness. If we are interested in the bearer's well-being in the long run, however, we must analyse his wants carefully. Only a few of them are of importance for *long-term* well-being. But from this it does not follow that wants or desires (in general) and the satisfaction of them are not of crucial importance for the person's well-being. This I wish to say to mark the difference between Sen's and my position.

Sen chooses instead to introduce the concept of *valuing*. He says that only the realisation of what is valued by *A* can contribute to *A*'s well-being. Sen and I may perhaps agree on many substantial issues. The wants which *A* considers to be important may perhaps often coincide with what she values. However, Sen and I try to express quite different thoughts. The act of valuation is something quite different from possessing a want or desire, says Sen.

I agree that there are acts of valuation. Valuation is necessary for people's living a life worth living. Valuations influence our desires and wants. So far we agree. The trouble, however, starts at this point. There are many kinds of valuations. Sen partly admits this when he acknowledges the person who chooses to send a substantial amount of money to Rwanda. There are moral valuations (and these are normally other-regarding), there are aesthetic valuations, intellectual valuations as well as valuations concering personal welfare. Only the latter valuations are automatically relevant in our context. The satisfaction of a moral valuation may contribute to the subject's well-being, if she is a well-integrated person and has incorporated these valuations among her personal wants. But, as Sen himself observes, choosing to act in accordance with one's ethics need not raise the level of one's well-being at all.

The reference to valuation in general cannot help us to understand or define the criteria for well-being. We must know first which are the valuations of welfare or well-being. But for that we need a definition of welfare or well-being. Hence, we are begging the question and going in full circle.

One can wonder how Sen in the end, when he is to specify the conditions for belonging to the evaluative space of well-being, can avoid relating his functionings and capabilities to certain ends, or the probability of achieving certain ends, such as happiness. If he cannot avoid such a reference, he and I will end up in a similar position.

Conclusion

In this chapter I have attempted to relate my theories of health and happiness to Amartya Sen's much-debated theory of well-being. I have looked at similarities and differences between our respective viewpoints. I have found that health, as I understand the concept, is one of many elements in what Sen calls his evaluative space of well-being. This is hardly a worrying conclusion. Health, also understood in the traditional medical way, as absence of disease, would qualify as a functioning in Sen's system and be an element in a person's well-being.

Our theories of well-being are quite different. Sen is a midfare-theorist. I am a final-end theorist. Sen has criticised the validity of my kind of theory. I have tried to answer these criticisms and at the same time questioned the possibility of Sen's clarifying what he means by well-being without referring to some kind of end-state, for instance happiness.

CHAPTER TEN

ON THE MULTIPLICITY OF HEALTH CONCEPTS

INTRODUCTION

There is a common complaint that there are so many different concepts of health and so many definitions of health that the enterprise of characterising health becomes completely hopeless. The question can then be asked: have I just contributed to the prevalent confusion by adding a further concept of health?

I hope not. In this chapter, moreover, I shall try to disentangle various sources of confusion in the area. I shall among other things ask the following questions:

a. In what sense are there a multitude of concepts of health?
b. What should we require of a reasonable concept of health?
c. Is it desirable to have a multiplicity of accepted concepts of health?

IN WHAT SENSE ARE THERE A MULTITUDE OF CONCEPTS OF HEALTH?

First, it is obvious that there are a great number of *proposed definitions* of health, from antiquity to the present day. To say that these all represent (or characterise) different concepts is, however, to make a superficial and misleading statement. Some of the definitions may (insofar as they attempt to characterise existing concepts) be obviously wrong. They may be contradictory or they may in other ways falsely represent existing concepts. They may in ways well-known in the theory of definition be too wide or too narrow or both too wide and too narrow.

What, then, is the criterion for the existence of a concept? This is a deep philosophical question to which I shall here only give a brief but for the purpose hopefully satisfactory answer. I shall accept two, each of them sufficient, criteria for the existence of a concept.

(i) There is a reasonably consistent use of a term connoting a concept. By this I mean that there is a linguistic community which continuously uses a term, say "health", in a coherent way. This is the standard criterion for the existence of a concept.

(ii) There is a serious and coherent proposal for the use of a new or previously existing term, and it is explicitly said that this is proposed as a stipulation. (I shall return to this case below.)

Consider first the case of a linguistic community. If there is such a linguistic community there is a use of a term and thereby a concept to describe. If one undertakes this enterprise one attempts to give a *lexical definition* of a concept. This lexical definition can be either correct or wrong.

Consider now some complications to this.

(i) *There is more than one linguistic community*. There are obviously numerous linguistic communities. In the most salient way the different languages symbolise such different communities. It is also common knowledge that terms such as "health", "Gesundheit", "santé" and "hälsa" are not exact synonyms. In a way, then, they represent slightly different concepts of health.

Even within a culture held together by a common language one can find subcultures having slightly or sometimes even radically different uses of the same term. Some of these subcultures are *technical or specialised cultures*. In the case of health the relevant subculture is the one of health care.

(ii) *There are various uses of the same term within one and the same linguistic community*. It is possible and even plausible that the members of one and the same linguistic community use a term in slightly different ways and that there is therefore no single neat concept to define. The different uses may even be contradictory. Take an example. There seems to be one (at least one technical medical) use of the term "health" according to which one cannot say that a person who has a disease can be completely healthy. On the other hand, there seems also to be a use according to which a person can be completely healthy in spite of having a disease.

Observe that I am here not talking about personal and idiosyncratic uses. I am talking about common uses and about possible cases of one and the same person taking part in contradictory practices.

I grant the validity of these observations. I also grant that these facts entail complications for a project of conceptual analysis. Moreover, I think it is a very important socio-linguistic project to try to map such diversities so that we can get a realistic understanding of their scope.

We need not, however, wait for the completion of such projects in order to do useful things in the clarification of concepts. A reason for some optimism is the following:

Even if members of one and the same community have slightly different uses of the term "health", they can still communicate with each other in using it. Some of these members can even take part in common scientific enterprises where this and related terms are crucial for communication. We have no feeling that this communication is completely ineffective. (We certainly know that it is not perfect and that there are cases where it does break down.) Most messages concerning health are reasonably effectively understood without pressupposing any previous clarification of language.

A plausible explanation of this fact of communication is that the various uses of the term "health" are not totally different but only partially different. The ideal situation (which is in fact a hypothesis explored in this and my previous essays) is that there is *one core* common to all or almost all of these notions. It is certainly a philosopically interesting project to find such a core and from this core try to develop a scientifically useful concept.

The scholar who has adopted the later Wittgenstein's [110] philosophy of language would not be impressed by this hypothesis. To her the directly opposite supposition would be the most natural: There is no *core* common to all uses of a particular term. The uses are linked to each other as members of a family. In such a relationship it is very likely that not all members have one feature in common. Nevertheless, it may be true that for every member there is some other member with whom she shares some characteristic.

I cannot in any abstract way resolve the dispute between the Wittgensteinians and myself. If my own analysis of health proves to be efficient this may, however, provide some evidence in this particular case.

Conclusion. The fact that there are many definitions of health does not entail that there are equally many concepts of health. A particular definition may be contradictory or otherwise inadequate as an attempt to characterise a concept. On the other hand I accept that there are a multitude of concepts in the following senses.

(i) There are many linguistic communities using the term "health" or one or other of its quasi-synonyms in English or other languages. These uses are not identical. Some of them may even be considerably different from each other. It is, however, plausible to assume that there is a core element in the vast majority of these uses. This core element could perhaps be called *the* notion of health.

(ii) It is obvious that there is more than one notion of health in the sense that there are many non-contradictory stipulative definitions of health. Such definitions cannot be criticised for inadequacy, since they are not intended to characterise anything at all. They are instead proposals for new discourses about health matters.

Definitions arising out of stipulation can, however, be criticised from other angles. They can be criticised from the point of view of *fruitfulness* or *reasonableness*. These are the criteria which I shall now consider.

WHAT SHOULD WE REQUIRE OF A REASONABLE CONCEPT OF HEALTH?

It is, as I have noted, indeed possible to define "health" in a stipulative way and introduce a new concept of health. The question is only what use one is going to make of such a concept. If it were completely unrelated to any known concepts of health it would hardly have any future use and would therefore be of no interest to us. As I see it, a stipulative definition of health would have to fulfil some subset of the following requirements.

(i) There should be some minimal conceptual relation between the introduced notion and the core-concept of health as characterised above.
(ii) There should be some justification for stipulations of the following kind.

1. The notion is justified for the simplification of certain bureaucratic decisions (for instance, in the social insurance system)
2. The notion is justified for the efficiency and/or simplification of communication which follows from using it in health care or in the health sciences
3. The notion is justified for its fruitfulness in scientific theory, either in medicine or in one or more of the other health sciences.[42]

The restrictions mentioned here entail that, in this context, I accept only such stipulations as are *explications* of the concept of health, i.e. as *on the basis of a given concept of health* propose such changes and simplifications as are warranted for well-defined purposes.

IS IT DESIRABLE TO HAVE A MULTIPLICITY OF ACCEPTED CONCEPTS OF HEALTH?

Can one system of thought embrace more than one concept of health? It is one thing that there is a variety of proposals of notions of health. But can one person or one linguistic community consistently live with such a variety? Would that not have a breakdown in communication as a consequence?

I think there are reasons for caution here. Any twofold use of a term involves a risk of misunderstanding and, unless the contexts of application for the two uses of the term are quite different, one ought to introduce a different term to mark the distinction. The conditions for stipulated pragmatic concepts must be very salient. For instance, it must be quite clear when we use the limited social insurance concept of health — which requires only that the person has a capacity for work in order to be healthy — and when we use the fullblown ordinary concept. A multiplicity of concepts designated by a single term is therefore never a desirable state per se.

A SPECIAL CASE OF HAVING TWO CONCEPTS OF HEALTH: BELONGING TO ONE LINGUISTIC COMMUNITY AND STIPULATING FOR ANOTHER

Let me now raise a further theoretical issue. One can share or possess a concept in at least two very different ways. First, one can subscribe to a learned definition of a concept. This is what we do when we officially side with a particular formulated definition of health. We say that this is *the correct* definition of health and we try to draw a number of conclusions from that. The second, and in fact the much deeper, way in which we share a concept is *in speaking and acting* in accordance with this concept. This way is deeper in the sense that the aim of a definition (normally) is to capture how we use a

42 Here I have only given an abstract presentation of such justifications. In every case we must then provide criteria for simplicity, efficiency and fruitfulness. In [71], pp.4-6, I have proposed a list of criteria for the fruitfulness of a scientific theory of health.

concept in practice. This is very much emphasised in the linguistic analytical tradition in philosophy, where use is considered a surer guide than received definition.

It is now easy to see that there may be a conflict between a person *A*'s possessing a concept in sense 1 and sense 2. *A* may side with a particular definition of health without realising that she, in her daily praxis, behaves as if some other definition is a much more reasonable one. This means that *A* has not reflected enough. She has performed a shallow analysis of her own praxis. Another alternative of course is that *A* wishes to make a change, wishes to introduce, by stipulation, some new definition and is ready to take the consequences of it.

An even more plausible case is that *A* may side with a particular definition but only partly follow it in her praxis. In fact, I should like to argue that this is the case with the medical establishment in Western countries. In many ways (in textbooks and in education in general) it has sided with something like a biostatistical definition. This seems to be quite obvious in pathology, laboratory medicine and some other highly specialised disciplines of somatic medicine, where diseases are defined and detected with methods which are quite remote from any considerations of human social functioning. However, this inclination is much less obvious and perhaps indeed not at all the case in certain instances of general practice and rehabilitation. Here the goal of health care in terms of the person's wellbeing and capabilities in life must of necessity be very much in focus.

In fact there are good arguments for saying that health care in the Western countries has a double platform, but that this is not made explicit and not well analysed in its own self-reflection.

We must thus distinguish between two ways of possessing a concept. First, we use the expression in the simple sense of "siding with a definition"; second, we use it in the important sense of "acting and speaking according to a concept". Siding with a definition can, but need not, change a lifestyle and a manner of speech. We can stipulate and thereby change our policies. Proposing a formal definition can, however, also leave praxis without any change at all.

On the Multiplicity of Health Concepts within the Framework of the Vital Goal Theory. Appearance and Reality

A discussion about the multiplicity of health concepts is not complete if we do not also consider *apparent* cases of such multiplicity but where this is ultimately not the case. I can illustrate this possibility by using my own proposed definition as a platform. The appearance of multiplicity is here dependent on the relational nature of the concept analysed.

Note again the formal structure of the concept of health introduced in this volume:

> *A* is in a bodily and mental state which is such that, given a set of accepted circumstances, *A* has the second-order ability to realise all her vital goals.

According to this characterisation health denotes a relation of equilibrium between a person's ability and her vital goals given a certain set of circumstances. Now, this equilibrium can be concretely realised in very different ways. First, the choice of accepted circumstances is crucial. "Running one's household" entails in a country like Sweden the execution of rather simple operations from a bio-mechanical point of view (walking to the local shop, placing saucepans on a stove and turning knobs on the stove), but it entails some rather advanced knowledge about how to operate the machines in the kitchen. To be able to run one's household, given the standard environment of Sweden, means to be able to handle all those things. "Running one's household" among Eskimos in Greenland entails something entirely different. The theoretical knowledge needed is different, and the manual skill needed is much greater and much more varied. Among other things survival there entails hunting and killing wild animals and the preparation of food from the stage where an animal has just been killed.

Second, the nature of the vital goals is very obviously crucial. If it is a vital goal for Smith to take part in the Olympic games, then this requires some significant athletic abilities on the part of Smith. If, on the other hand, it is a vital goal for Jones to reach a top position in the Conservative Party, this requires some very different abilities on the part of Jones.

We see, then, that the health of the Swede and the health of the Eskimo, as well as the health of Smith and the health of Jones, can entail quite different ability configurations, depending on the variability of circumstances and goals. This is a simple logical conclusion from the fact that the concept explicated is a relational concept.

This variety, constituted by the various instances of health, does not however entail that the term "health" is ambiguous or that there are two concepts of health in operation. The variety of the terms of a relation does not entail the ambiguity of the relation itself.

ON SOME VARIANTS OF HEALTH CONCEPTS WITHIN THE FRAMEWORK OF THE VITAL GOAL THEORY

I have just discussed a case where there is an apparent multiplicity of health concepts. The appearance was due to the fact that the basic concept is relational, which may in some instances entail that two people with different biological make-ups can have the same degree of health, and conversely two people with the same biological make-up can have different degrees of health.

There is, however, a sense in which one can talk about a multiplicity of health concepts within the general framework of the vital goal theory. I shall here distinguish between two ways in which such a multiplicity can be formed.

(i) *Variants of health concepts formed by relaxing some of the terms in the relation*

Our basic notion of health says that *A* is healthy when she is able to realise all her vital goals given an accepted set of circumstances. We can create variants of this concept just by slightly changing these conditions. I shall discuss two such plausible strategies.

According to one such strategy, in fact already suggested, it is plausible to distinguish between *complete* health and all the conceivable stages to a point which might be called *minimal* health. Below this point there is, as we say, no positive health, but instead illness. In order to have minimal health the subject need not of course have the ability to realise all her vital goals but only a subset of them. Thus we can create a great number of variants of the concept of health by distinguishing between different sets and subsets of vital goals.

Other variants can be distinguished by looking at the circumstance-term of the relation. So far we have more closely considered the ideas of *accepted circumstances* and *standard circumstances (given a certain culture)* as the basic ideas behind our ordinary notion of health. It is, however, plausible to consider other sets of circumstances for particular purposes.

There is, it may be contended, a stronger notion of health than our ordinary one which is less dependent on a particular culture-relative set of environments. Compare *A* who is completely healthy given her standard environment *E* (and no other environment) with *B* who can fulfil her vital goals not only in her standard environment *E* but also in a great number of other environments *E1...En*. According to the stronger notion of health *B* is here healthier than *A*. We might call this the *adaptability* notion of health. It indicates that the healthy person has greater adaptability to new kinds of situations than the less healthy one. (Observe that adaptability is a property that is necessary for health even given a set of accepted circumstances. This set is to be explicit rather a range of sets. The accepted circumstances may vary within certain limits. Moderately harsh circumstances, as well as moderately advantageous circumstances, are accepted within these limits.)

Adaptability does not merely depend on the person's psychophysical strength. Adaptability is also a matter of *realism*: to what extent is the subject prepared to change her ambitions (i.e. some of her vital goals), to reformulate them in the light of a new environment? If the subject is adaptable in this sense some lack of psychophysical strength can be compensated for and the equilibrium between ability and vital goals can remain.

(ii) *Variants of health concepts depending on the causal background of health*

The concept of health in this volume is in a way a very general concept. It is in particular *not* an exclusively medical concept. I have so far said nothing about the relation between health and the more clearly medical concepts of *disease, injury and defect.*

To claim that *A* is ill, given our conception so far, is to say that *A*'s ability to realise her vital goals is compromised by *some* internal cause, so far unspecified.

It is now possible to construct more specific and narrow concepts of ill-health or illness having the causal background of the illness as a ground for division. Some (perhaps the majority of) instances of illness are caused by a *disease*, i.e. an internal bodily or mental process that typically results in illness. Other instances are caused by *injuries* or *defects*. Both can be viewed as internal states of affairs, the former acquired, the latter congenital, which typically result in illness.[43]

---

43 For a more elaborate discussion of these concepts, see [71], pp. 108-109. Observe that I am inclined to view impairments as similar to injuries in the mentioned respects. Perhaps impairments can be seen as a

Together these three concepts form what are often called *maladies*. Illness caused by any of the three kinds of entities might then be called a malady-form of illness or just a *medical* form of illness.

Given the adjective *medical* we may then also construct new positive variants of health. We may say that a person is *medically healthy,* if and only if her health has not been compromised by any of the malady-forms, disease, injury (impairment) or defect. Likewise we may make a further subcategorisation and say that a person is physically healthy if her health has not been compromised by physical diseases, injuries or defects, and that she is mentally healthy if her health has not been compromised by mental diseases, injuries or defects. Physical health is then compatible with mental illness and vice versa. Moreover, medical health in general is compatible with ill-health due to other causes. A person may be medically healthy but still ill because of, for example, existential problems.

---

subspecies of injuries, viz. such injuries as are caused by disease. See the conceptualisation suggested in chapter 12.

PART III

Action-Theoretic Applications in the Theory of Health and Health Care

CHAPTER ELEVEN

ON THE TECHNICAL NOTIONS OF DISABILITY AND HANDICAP: THE WHO CONTEXT

INTRODUCTION

In this section I shall comment on and attempt to characterise the technical notions of disability and handicap that have been proposed by the WHO mainly through their document *International Classification of Impairments, Disabilities and Handicaps (ICIDH)* [41]. Although the main purpose of this document is to provide a nomenclature for the classification of species of impairments, disabilities and handicaps, it also contains very serious attempts to define the fundamental generic concepts. I shall critically examine these definitions and, in particular, focus on the role that the notion of *normal ability* plays in them. I shall also examine the alternatives to these definitions proposed by the Canadian and Quebec committees on the ICIDH.[44]

In this critical discussion I shall partly be agreeing with the suggestions developed by the Canadian and Quebec committees on the ICIDH. I endorse their dismissal of the concept of normality from the definition of handicaps, and I agree with their emphasis on the environment as a constitutive part of a handicap.

For the assessment of these proposals I shall use my previous analysis of the basic concept of *ability*. According to this analysis ability is not a simple predicate but a *three-place relation*, where the terms are: the *agent*, the *goal of an action* and the *circumstances* surrounding this action. Disabilities and handicaps are in an analogous way to be construed as three-place relations. When one considers the three-place nature of disabilities and handicaps one realises that for every disability and handicap there is, apart from the agent, a set of goals and a set of circumstances to be made explicit. If these are not defined in the characterisation of a particular disability or handicap the characterisation is, strictly speaking, empty.

The choice of the goal term in the definition of disabilities and handicaps is crucial and has both ethical and political dimensions. In my discussion of this issue I shall use the notion of a *vital goal* which has been characterised above. My general suggestion then is that *disabilities and handicaps should be determined in relation to the individual's own vital goals*.

I argue that disabilities and handicaps are alike in their basic logic. I also argue that the attempts in the ICIDH discussion to distinguish between them have not been convincing. There are, however, still good reasons for separating a category of disabilities from the category of handicaps. I propose a method for such a separation using the distinction, introduced above, between *basic actions* and *generated actions*.

---

44 For a detailed analysis of these definitions, see my [66]. For critical remarks from an empirical standpoint, see E.M. Badley [7], and M.R. Bury [15].

As a summary of my analysis I propose two schemata for the definition of specific disabilities and handicaps.

On abilities as three-place relations

As I have already noted, ability is not an absolute concept. It is pointless to say of a person that she is in general able, or conversely, that she is in general disabled. Ability has to be specified. First, one has to identify a particular agent *A*. Second, one has to specify a project or goal of *A*'s, i.e. something that *A* is able to attain; in many cases this goal can be identified with a simple action, i.e. *A* is able to perform this action. And third, one has to specify the circumstances in which *A* is able to attain this goal or perform this action.

Thus ability is a three-place relation with the following terms:

the agent involved

the goal of this agent

the circumstances in which the agent acts.

The difference between ability and disability can be expressed by the following statements: When *A* is able to reach *G* in *C*, then *A* reaches *G* in *C* if *A* tries. When *A* is unable to reach *G* in *C*, then *A* does not reach *G* in *C* if *A* tries.

We can now also see how we can use this schema for analysing all kinds of ability- and disability-relations. We can for instance see how we can express extreme cases of ability and disability. Mr. Smith, for instance, may be unable to do his own cooking in *any* kind of circumstances. He is then *completely* disabled with respect to this action.

On some medical characterisations of disabilities and handicaps and the accompanying proposals concerning the goal term of ability

Using this piece of analysis as a background let me now turn to the ICIDH and its proposed definitions of disabilities and handicaps. Let me first as a background notice that the ICIDH is founded on a general hypothesis — or rather a presupposition — to the effect that there is a normal causal sequence of pathological states in the following order:

> *Diseases* cause *impairments*, which in their turn cause *disabilities*, which in their turn cause *handicaps*.

A disease, according to the ICIDH, is a state which causes a loss or abnormality of psychological, physiological or anatomical structure or function (i.e. an impairment); an impairment causes a restriction or lack of ability to perform an activity in the manner or within the range considered normal for a human being (i.e. a disability); a disability, finally, causes or may cause a disadvantage for a given individual which limits or prevents the fulfilment of a role that is normal for that individual (i.e. a handicap).[45]

I have hereby implicitly presented the definitions of the central concepts as presented in the ICIDH. Impairments are losses or abnormalities of bodily structure or function. Disabilities are restrictions of ability to perform activities. Handicaps, finally, are disadvantages that limit or prevent the fulfilment of a role.

We can now ask: how does the ICIDH handle the three-place nature of disabilities and handicaps? Do the authors of the document agree that such is the nature of disabilities and handicaps? To the latter question we can answer: they partially agree, but it is not shown consistently and clearly. In order to answer our first question we therefore have to do some interpretation and reconstruction.

The agent- or person-term is of course obvious. No-one can neglect that. Disabilities and handicaps are properties related to individual human beings. But what about the goals? Which are the goals, according to the ICIDH, that a disabled or handicapped person cannot reach? Or, in other terms, which is the set of actions that she cannot perform? And in the event that we get a clear answer to this question, why do we select this particular set of goals or actions?

The ICIDH does not have explicit answers to these questions. We have to interpret the document. The ICIDH does not offer any list of goals or actions which are such that an able person should attain them if she tries. It answers instead in an indirect way. It says: a disability is a lack of ability to perform an activity within the range considered *normal* for a human being. A handicap is a disadvantage preventing the fulfilment of a role that is *normal* for that individual.

In short, a disabled or handicapped person is unable to perform an activity or fulfil a role that is considered to be normal. The concept of normality thus plays a major role in the ICIDH, but it is not analysed in the document at all. This is highly regrettable, since the interpretation of normality is completely crucial for the application of the relevant concepts.

Normality has been discussed in many places in the philosophy of medicine, not least in the context of health, and it has been noted that there are many different interpretations of this concept. For my purposes it is sufficient to take a short-cut. The main distinction that I wish to make is between a *statistical-descriptive* interpretation of normality and a *normative* interpretation. According to the former such an ability is normal as falls within a range that covers the abilities of the majority of the population in the relevant respect. Thus, if most people can lift weights of at least 100 lbs, then a particular person is normal in this respect if she can do so, and abnormal if she cannot.

According to the latter interpretation an ability to lift such weights is normal, if and only if it fulfils a norm which says: an adult human being should be able to lift weights of at least 100 lbs.

There are many signs in the text of the ICIDH which imply the exclusion of the statistical interpretation. One such sign is the phrasing of the definitions, where the author writes about an ability *considered* to be normal or about the fulfilment of a role *consid-*

45 See the *ICIDH* [41], pp. 25-30.

ered to be normal. If an ability is considered to be normal or abnormal it is unlikely that this consideration is based on a statistical calculation. The most natural interpretation of the phrase is that it refers to some cultural norm of roughly the following kind: if you are a member of British or Swedish society you should be able to do such and such. This norm may contain the implicit assumption: if you cannot, you will not be completely at home in these societies.

If the interpretation that I am just making is sound then it is true that the ICIDH, and perhaps by implication also the WHO, advances a paternalistic interpretation of the goal-term in our analysis of the ability-concept. A person who is able in a particular respect or in general should be able to attain or accomplish what is considered to be normal in the society in which she happens to reside.

This is *not* a self-evident interpretation of the goal-term. There is no logical reason that compels us to assume that a person is able only if she can do what is considered to be normal. As a matter of fact many sensible people have taken issue with this assumption. There are influential critics of this view even among commentators who work in the same tradition as the ICIDH. I shall here particularly refer to the Canadian Society for the ICIDH and the Quebec Committee on the ICIDH [16]. (In the following I shall refer to them as the Canadian committees.)

The Canadian committees have taken issue with the normative idea of normality, at least in the context of *handicaps*. They say that we should not define handicaps in terms of vague cultural norms. What we should look into is *the individual herself* and her specific *life habits*. The idea then is that the individual is handicapped, if and only if she is unable to perform one or more such actions as are included in one such habit. If it is included among the habits of Mr. Jones or Mr. Smith to lift weights, or to lift heavy objects in general, and Jones or Smith at a particular time happens not to be able to realise his habit, then at this particular time he is handicapped.

If, on the other hand, Jones and Smith are professors of philosophy, who never lift any heavy physical objects at all, or at least for whom it is not a habit to lift heavy physical objects, they will never be handicapped in this respect. And this is so whatever their basic abilities are. A professor of philosophy who can only lift a weight of 25 lbs is not handicapped, since he has probably no use for lifting more than 25 lbs.

I am much in favour of the Canadian idea of relating handicaps to the individual and her habits. I think that this is a more plausible suggestion both from a moral and from a conceptual point of view. The culture or the society should not and need not (except in very particular cases) impose any norms on the individual in the context of health care. Moreover, it is much more likely that we should get a clear answer from the handicapped individual about what her life habits are, than that we should get a clear answer from the society or any of its parts about what its norms for a particular ability are.

I think, however, that the Canadians have not found the best possible answer. First, it is in many cases simply impossible to use a person's life-habits as a criterion for health or ability. This is most easily seen in the case of genetic handicaps or disabilities. A person with a grave handicap of orientation, for example, which is due to a genetic visual defect, has since birth been forced to form her life-habits in accordance with her handicap. Thus, the life-habits themselves involve the handicap. As a consequence, the handicap cannot be detected as a deviance from or an interruption of the person's life-habits. Presumably, however, the Canadian group wish to include genetically determined disabilities and handicaps in their classification.

Secondly, why should the criterion merely refer to habits? This restricts us to considering repeated or repeatable actions only. But a person's life is not solely defined by her habits. People have many particular goals outside anything that could be called habits of theirs — goals which they wish to realise. A person may have the goal to travel to a particular place, or to take an exam, or to perform some other specific act. If these goals are very important to the subject, they belong to what I have earlier called her *vital goals*.

At a particular moment the individual may not be able to perform an action which is a necessary condition for realising one of her vital goals. Although the action is not often repeated it may be of crucial importance to the individual. The institution of health care therefore ought to notice this case, at least if the non-ability on the part of the individual is of the kind that standard health care can do something about. Thus I would say that the individual is handicapped, although the action in question is not included among the actions which she performs habitually. Hereby I suggest a much more radical dismissal of the concept of normality from the arena of handicaps.

Now I have to answer an argument that could be raised against both the Canadian suggestion and my own. The argument runs as follows. Is there not a danger in relating the notion of ability to the agent's own habits or goals? Could we not in some cases come up with counterintuitive results? Suppose that a person has very poor or even completely detrimental habits. Suppose that she is an alcoholic who does little beyond what is necessary for bare survival except go out to get more liquor. Would it not be rather strange to judge this person's abilities only in relation to the habit that she happens to have, and which in this case is completely detrimental? And could we not devise a similar argument, *mutatis mutandis*, against my idea of relating abilities to a person's vital goals?

I can here only defend my own case. As my analysis above has shown, the notion of vital goals is not equivalent to any state of affairs that a person wants or intends to become the case. If it were I would be a clear target for the objection in question. According to my concept, a person may want, indeed consciously want, to attain something which is not a vital goal for her. This is obviously so in the case of alcoholics and drug-addicts.

ON DISABILITIES, HANDICAPS AND CIRCUMSTANCES

So much for the first and second terms in the three-place relation of disability and handicap. Let us turn now to the third term, the one that concerns the circumstances under which an ability can be exercised. What does the ICIDH say here? There is an explicit acknowledgement. The ICIDH notices, at least in the case of handicaps, that the realisation of a role depends not only on a basic personal ability but also on the external situation in which the individual is placed.

When, for instance, a person manages to do her professional work properly, there is always a context within which she does so. The person's work-place can be more or less practical, the person's psycho-social environment can be more or less favourable, and her success is obviously to a great extent dependent on the structure and content of this environment. The Canadian group is even more explicit about the fact of context-dependence; they underline it and make it a matter of policy in their definition of handicaps. A handicap, according to the Canadian document [16], is the disruption of an individual's performance of life-habits — taking into account age, sex and socio-cultural identity — resulting, on the one hand, from impairment or disability, or, on the other, from obstacles attributable to environmental factors.

Thus a handicap can be created both via damage to a person's body or mind, and via the existence or introduction of one or more preventive factors. All of us can be, and indeed are, handicapped with respect to certain goals because of the way the world is. As a Swede I am most of the year handicapped with respect to sunbathing. The explanation is not that I have been injured or am ill in any way. It is simply that there is very little sun in Sweden and thus very little opportunity for sunbathing.

Moreover, it is easy to see that all of us can become handicapped in any respect if a negative change occurs in our environment. Nature can become so rough that we can hardly do anything at all. We can encounter a hurricane or an earthquake. And society can become so complicated or contradictory that certain members of it can hardly accomplish anything useful at all. The latter is unfortunately a fact to many people in many modern societies.

When a handicap has been created more or less exclusively by external change we often use the term "social handicap". Many medical definitions and nomenclatures therefore tend to ignore and avoid such a handicap. A handicap proper, according to the reasoning of both the ICIDH and the Canadian documents, is a disadvantage which is at least partly due to a disease or an impairment afflicting the subject.

I think that there is a danger in using the label "social handicap", in particular if it is used as a reason for dismissing such handicaps from a medical discussion. One important reason for seriously considering all kinds of handicaps in a medical context is the relativity of the concept of disease itself. Disease-nomenclatures have come and gone, and they are to some extent culture-dependent. And in spite of strong international efforts to create a common system of concepts there is no worldwide consensus about the status of categories such as alcoholism, homosexuality or psychopathology. It seems to be odd that we must first have an official recognition of the existence of a disease or an impairment in a subject before we are allowed to use the term "handicap" in a characterisation of the subject or of her situation.

Having observed this it is still important, not least for practical purposes, to be able to distinguish between such handicaps as are mainly attributable to external change and such as are attributable to internal processes. What I am taking issue with is the idea that there is a clear, universal and eternal rule for establishing when a handicap is medical, i.e. internal, and when it is social, i.e. external. There are instead different rules varying from one time to another and from one culture to another. The rule to be used must be defined and agreed upon in every context of discussion. The involved parties have to agree upon what is to be counted as the *accepted* context or the *accepted* circumstances. This means that they have to agree upon what is to be counted as the type of circumstances in which a subject should be able to act in order to be looked upon as able in a relevant respect, and which cannot be counted as preventive circumstances.

Let me give a concrete example: What should be the accepted set of circumstances for the activity of walking? What temperature, what strength of wind and what kind of surface should one be able to manage in order to be qualified as a walker? The span of what we today in Sweden consider to be standard circumstances for walking is great. A walker should be able to manage almost any temperature, she should be able to stand a wind of at least 20 metres per second and she should be able to manage a great variety of rough circumstances. Circumstances within this defined set cannot be viewed as preventive. Thus if a person is unable to walk, given these circumstances, her handicap or disability must be of an internal or, if we wish, medical kind.

When I say that the involved parties must *agree* upon a set of circumstances in a particular context, I am not only making a logical but also a *political* point. The distinction between internal and external handicaps or disabilities presupposes *some* definition of standard circumstances, but it does not presuppose an agreement. It is logically possible that the standard circumstances are proclaimed and defined by *authorities*, for instance medical authorities. And this is probably what has mostly been the case in the past. If the whole conceptual issue that I am now exploring is not made explicit (which indeed is very often the case) then there is no opportunity for negotiation. Hence the person who has the power of decision, a doctor or some other kind of person in a position of authority, is also the most likely party to define whether a medical or a social handicap is present or not.

In order not to be misunderstood on this matter I shall emphasise a distinction here. I have throughout been talking about the set of circumstances in relation to which a person shall be judged to be able or handicapped in a certain respect. This set which has to do with the very definition of handicap must not be confused with the kinds of circumstances that we require that the society shall promote in order to *compensate* for handicaps, to help people who are handicapped in relation to the particular society. This is of course the most obvious political issue. My present point is different but it still has a political dimension. It is the point that the notion of handicap cannot be a completely theoretical, for instance biological, notion. The very identification of a person as handicapped presupposes an ethical or political decision.

ON THE NOTION OF DISABILITY

So far I have for the most part used the term "handicap" and avoided the related term "disability". Moreover I have explicitly related what I have said to the characterisation of handicap which was suggested in the ICIDH and in the Canadian ICIDH texts. But these documents also discuss the notion of disability, and there is a set of interesting ideas expressed in the ICIDH also about disabilities. I shall now turn to consider this notion.

According to the ICIDH (and here the Canadians seem to share the principal ideas, although they partly draw different conclusions) a disability is characterised by the following:

a. it is a restriction or lack of ability to perform an activity or a task (in contrast to fulfilling a more complex role)

b. the lack of ability is completely or almost completely context-independent

c. the lack of ability is considered to be abnormal (or to put it more cautiously: transcends a range of behaviours which are considered to be normal)

d. the identification of a disability can be made from a scientific, neutral, standpoint, whereas the identification of a handicap requires some kind of evaluation.

In the following I shall explicitly take issue with points b and c. From this follows that I disagree with d. Instead I shall propose a criterion to distinguish between handicaps and disabilities which resembles the one found in a.

In particular the Canadian committees stress the fact that a disability, in contrast to a handicap, can be *independent* of the environment. In short, a disability is the case whatever the circumstances are. This dictum of course completely contradicts my initial analysis in this paper which says that every case of non-ability is a case of a three-place relation, involving the terms: an agent, a set of goals and a set of circumstances.

I shall here defend my point and say that the idea of context-independence is misleading if not a mistake. I doubt that there can be such a concept of disability as the Canadians assume.

There is a prima facie plausibility in their dictum. By the term "disability", as I have indicated above, they intend mainly to refer to basic bodily functions or bodily movements. The Canadian committees include in their list of disabilities such basic categories as physiological functions, sensory functions and motor functions. It may seem that the exercise of such functions is totally context-independent.

It is, however, an important logical point that all abilities, however simple and basic they are, are abilities in relation to a set of circumstances. To say that being able to move one's leg is an ability irrespective of a situation is to say something false. When one can move one's leg, one can move it given a set of circumstances which could in principle be specified. One can move it given the standard atmosphere of the earth, given a certain climate, excluding certain hurricanes and storms, given the non-existence of impeding physical objects and so on. Similarly, when one can hear sounds, one can do so given a certain specifiable set of circumstances excluding various disturbing or distracting noises.

To make this logical point is not just to say something sophistic or self-evident that one need not bother about in practical work. On the one hand, the circumstances which are often taken for granted may be absent in a particular case, so that a person very misleadingly could be deemed disabled. Assume, for instance, that a person's ability to hear sounds is tested in a room where there are some disturbing noises. She may then be unable to interpret or even detect the primary sounds. It would, however, be quite misleading to consider this person disabled, since she may be a perfectly competent listener in a room which is quiet or which does not exceed what we consider to be the standard maximum of background sounds. On the other hand, it may very well be the case that a certain person comes out on the able side simply because the measurement of her ability occurs when the circumstances are extremely favourable. For instance the testing of the ability to detect some sounds might occur in a situation where the sounds happen to be magnified by a microphone.

Such pieces of information are important not least in the discipline of rehabilitation. A physiotherapist or occupational therapist, in attempting to rehabilitate a patient, i.e. to enable her to do certain things, must know quite clearly under what circumstances this ability should be exercised. It may be the case that the therapist assumes a set of standard circumstances, e.g. the circumstances that are most often the case for the patient in question. But since standard circumstances may vary from individual to individual, they ought to be specified for the sake of clarity in medical communication. To realise that all abilities, however simple, are context-dependent is therefore of great importance also for the general purpose of health care.

So much for context-dependence in the case of disabilities. But what about the goal-term? What behaviour should one be able to perform given a specified set of standard circumstances? On this point both the ICIDH and the Canadian document [16] use the concept of normality to define the relevant class of behaviours. The Canadian definition runs as follows: "A disability is any reduction, resulting from an impairment, in the capacity to perform a physical or mental activity considered normal for a human being (depending on his biological characteristics)".[46]

But do they here really intend what they say? There are many movements which we could conceive of performing with our limbs or with our lips and tongue. It may be true that most of us have the ability to perform them. We may be able to move the little toe up and down at a certain angle. We may be able to produce the most interesting noises with our tongue and lips. But only a subset of such things is to be found in existing medical lists of abilities and corresponding disabilities. Why is that? Do they not fulfil the requirement of normality? How, then, is normality to be understood?

In analogy with what I said above about handicaps I don't think that the concept of normality is helpful here. My own explanation of the selection which is traditionally made is of a different kind. The reason why these behaviours and the corresponding abilities and disabilities are not included must be that they do not serve any useful purpose in the culture within which we happen to be living. If, for instance, the Swedish or English oral language had needed the particular movements of the tongue and lips, in the sense that they could play a role in the conventional linguistic code, then they would have been important. As a consequence they would also constitute abilities in an interesting sense.

But are, then, the requirements of the culture the determining criteria of what should count as abilities and disabilities? Are the determining criteria the ways institutions operate and the way they make demands on people? From a logical point of view this could be the case, and this is perhaps also what has often happened in practice. But as in the case of handicaps we are here faced with a political choice. Are we interested in the

---

46 See the *ICIDH International Network* [16], p. 27.

impersonal norms of the culture? Or are we interested in the individual's own *vital goals?*

For my own purposes the answer is easy. A consistent way of looking at disabilities and handicaps presupposes the second alternative. A disability, as well as a handicap, is a non-ability — given a specified set of circumstances — to realise one or more of one's vital goals (or any of its necessary conditions). The individual's own vital goals are certainly partly influenced by cultural norms and cultural demands, but they are not completely determined by them.

But is there, then, a rationale for distinguishing between handicaps and disabilities? So far I have only pointed out similarities. Yes, I still consider that there is an important line of demarcation. And the line which I shall suggest would single out roughly the same two classes as the criteria of the ICIDH and the Canadian document [16] attempt to single out.

Consider the distinction introduced in chapter 3 between *basic action*s and *generated action*s. A generated action is an action which is performed *by* the performance of some other action. Consider, for instance, the action of building a house. When one builds a house one does so *by doing* several other things, e.g. by putting pieces of wood together or by laying bricks on top of each other. These subordinate actions may in turn be generated. One lays bricks, for instance, by moving one's arm up and down. Such a chain of actions finally reaches rock bottom, i.e. it arrives at a kind of action which is not done by the performance of something else. Such an action is a *basic action*.

This pair of concepts can now be used for the following definitions: a disability is a non-ability to perform a basic action, for instance a non-ability to lift one's arm; a handicap is a non-ability to perform a generated action, for instance a non-ability to perform one's work properly.

A NOTE ON THE AGENT-RELATIVITY OF DISABILITIES AND HANDICAPS

Throughout this chapter I have advocated an individual-centred philosophy of disabilities and handicaps. I have suggested that we should relate our concepts of disabilities and handicaps to the *individual's* own vital goals. I propose that a person is disabled or handicapped with respect to a particular goal or action, if and only if this person is unable to realise this goal or action, given a set of circumstances which have been agreed upon.

Regarding this proposal the sceptic might wish to put the following questions: Are the implications of the individualised view of disabilities and handicaps, which has been presented in this chapter, tolerable? Does it imply that there is no general way of describing abilities and disabilities? How should we then at all be able to communicate on these matters in the scientific community?

I think that there is little reason to fear such negative consequences. To start with, people's sets of vital goals are not as distinct as my previous discussion may suggest. We all have, more or less for biological reasons, a number of vital goals in common. We all need to survive. Therefore most of us have to earn our living through work. As a further consequence we must all exist within a society with its typical institutions and norms. A very important consequence of this is that it is a vital goal for all (or almost all) people to be able to speak and otherwise use a language. This implies that impairments in the "language" organs, including damage to the ear and the eye, will lead to disabilities and handicaps for almost all people. Thus a nomenclature for such disabilities and handicaps can have an almost universal application.

It is true, in particular in the field of handicaps, that there may be great variations between different persons. People's jobs require different capacities; their leisure life may call for the use of varying abilities. This diversity, however, can hardly be a problem for international communication. It may be true that certain handicaps are rare, because of the rarity of the vital goals presupposed. The non-ability to take part in acrobatics is a handicap only to circus-artists. The non-ability to hide a rabbit in one's coat is a handicap only to a conjurer.

But the rarity of an entity does not generally prevent us from having a nomenclature in other fields, let alone the science and practice of medicine. There are a great many rare diseases which are given names in the international scientific classifications. Admittedly, there is a logical difference between, on the one hand, diseases and, on the other hand, disabilities and handicaps. But this is a difference that we must accept: in the case of disabilities and handicaps two persons may be completely alike from a bodily point of view. They may both be unable to do *F*. It may, however, very well be the case that only one of them is disabled or handicapped with respect to *F*.

CONCLUDING REMARKS

In this chapter I have presented a comprehensive view of disabilities and handicaps where I have indicated the common logic of these two concepts and investigated some of the consequences of this logic. Fundamentally I have underlined that both these notions are three-place notions involving the following terms: the agent, the goal of the action in question and the circumstances of the action. I have stressed that a particular characterisation of a disability or handicap must specify all these terms. I have also taken a particular stand concerning the goal-term of the relations of disabilities and handicaps. I have suggested that disabilities and handicaps should be related to the individual's own vital goals.

Although I claim that disabilities and handicaps are logically similar concepts, I think that there is a useful distinction to be made between them. I then follow the suggestion originally made in the ICIDH that disabilities relate to simple movements of the body or to simple tasks, whereas handicaps relate to more complex actions or roles. In an attempt to clarify this distinction I have borrowed the pair of concepts *basic action* and *generated action* from modern philosophical action theory.

Let me now as a summary present my definition schemata for the two central concepts:

A is *disabled* (with respect to action *F*), if and only if
(i) *F* is a basic action

(ii) *A* is unable to perform *F*, given an accepted set of circumstances which have been agreed upon in the context
(iii) *A*'s performance of *F* is a necessary condition for the realisation of one or more of *A*'s vital goals.

A is *handicapped* (with respect to action *F*), if and only if
(i) *F* is a generated action
(ii) *A* is unable to perform *F*, given an accepted set of circumstances which have been agreed upon in the context
(iii) *A*'s performance of *F* is a necessary condition for the realisation of one or more of *A*'s vital goals.

A final crucial question remains. What are the relations between the notions of health, disability and handicap? The characterising elements — for instance, vital goals and accepted circumstances — seem to be identical for the three notions. Does this mean that disability and handicap are simply variants of ill health?

My answer to this is that the three notions are strongly related to each other. My whole approach to the philosophy of health is consciously action-theoretic. A person's lack of ability to realise certain goals is a core element in my characterisation of health. However, it is wrong to say that my concept of health can be completely reduced to either disability or handicap. Let us recall the formal definition of health in chapter 9: A person is completely healthy, if and only if she is in *a bodily or mental state which is such that she has a second-order ability,* given accepted circumstances, to realise the states of affairs which are necessary and together sufficient for her minimal happiness in the long run. Thus first, health is not identified with the ability itself. Health is identified with a compound bodily and mental state. What makes this state healthy is its particular dispositional property of enabling its bearer to perform a set of crucial actions. And second, the ability, in the case of health, is of a second-order kind. The notions of disability and handicap, on the contrary, can be used also in the context of a first-order non-ability.[47]

[47] The distinction between disabilities and handicaps has recently been criticised by S.D. Edwards [27]. For a response see my [74].

CHAPTER TWELVE

ON THE COMPLEXITY OF AUTONOMY

INTRODUCTION

The term "autonomy" is now used frequently in ethical contexts. Beauchamp and Childress in their often cited *Principles of Biomedical Ethics* [8], for instance, refer to the principle of respect for autonomy as one of four fundamental principles in medical ethics. The general idea of respecting a person's autonomy is central in most ethical codes, be they research-ethical or professional ones.

The rough idea is easy enough to grasp. The theories and codes maintain that it is our duty to respect a person's right to govern her own life in certain basic respects. More specifically, the right normally concerns the person's right to decide what should happen to her in a particular situation. In the case of medical ethics it often concerns the patient's right to consent or not consent to a particular treatment or a particular health-promoting measure.

However, there are clear signs that the term "autonomy" can have rather different connotations. We now find texts which do not just talk about our duty to respect autonomy but also about the importance of *creating* autonomy. There is, for instance, a locution of roughly the following kind: It is sometimes right to infringe upon a person's autonomy in order to increase her future autonomy.[48]

What is the sense of "autonomy" in the latter occurrence of the word in this locution? Is it really the same as in the first occurrence? I think that the answer remains quite unclear. It shall therefore be my concern in this chapter to investigate this question.

AUTONOMY AS A THEORETICAL PROPERTY

The first necessary distinction to be made is the one between, on the one hand, autonomy as a theoretical property and, on the other hand, the *right* to have and execute this property. Roughly, to be autonomous in the former sense is to be, *de facto*, a self-governing person. This in turn can be split up into two cases. First, the sense of *having the power* to govern oneself. Second, the sense of actually *executing* this power. A person can clearly be able to govern herself without using this power in practice.

From this follows that the normative question, in its turn, must be split into two. A person can have a right to have a particular power, and the person can have a right to execute this power in a certain set of contexts.

48 See in particular D. Seedhouse [89] for the idea that the creation of autonomy is one of the fundamental goals of health care.

A further important distinction is the one between a person's *decision* to act and her actually acting in accordance with this decision.

Let me first investigate somewhat the following basic theoretical question. What is it to have the power to govern oneself? The analysis of this concept is, I shall suggest, partly identical with the analysis of the notion of a person's *practical possibility of acting*. The two basic notions in this analysis are, as I have argued above, *ability* and *opportunity*. The definition runs as follows:

> *A* has the practical possibility of doing *F*, if and only if *A* has both the ability and the opportunity to do *F*.

At least part of the sense of having the power to govern oneself is, I shall suggest, to have *both* the ability *and* the opportunity to govern oneself. What does this then entail? Let me begin with *opportunity*.

The power of self-government presupposes the non-existence of preventive factors. A person clearly does not have the power to run her affairs if there is something externally preventing the execution of this power. A person does not have the power to enter her work-place if there are pickets blocking the entrance. The preventive factor is in this case a number of human beings. But it could very well be a non-human factor, for instance a hurricane. (This analysis can be complicated further, if we take time and thereby changes in the external world into account. I shall here restrict myself to the notion of a power in a particular instance.)

So far I have restricted myself to prevention of physical action. But also mental actions such as decisions may be externally prevented. A person may be unable to decide on a particular matter because she is continually disturbed or has too little time to reflect on the matter.

Thus we have a basic sense of autonomy or power of self-government: *freedom from external prevention in deciding and acting*.

There is also, however, a sense in which both decision and physical action can be *internally prevented*. I have in mind, for instance, interference by alcohol and drugs or indeed by the abstinence from them. The agent in such a case, we would say, does not have the opportunity for free deliberation.

The notion of internal prevention is in principle different from, although sometimes difficult to distinguish in practice from, *ability* itself. By ability I mean roughly the set of bodily and mental conditions of the person which together with a set of external circumstances constitute full power. (A way of making the distinction is the following. Factors such as poison or drugs, which are not actually parts of the body but operate inside the body, are instances of internally preventive factors. Disease-processes and injuries, however, are factors which contribute to the constitution of the disability itself.)[49]

As I pointed out in the initial analysis of practical possibility, ability is always an ability in relation to a particular set of circumstances, i.e. in relation to a particular opportunity. The two concepts must therefore be discussed jointly. There is no such thing

[49] The issue of internal and external conditions for ability is further discussed in my [73].

as an ability *in vacuo*. When we say that a person is able to do something, we must always tacitly assume a particular set of circumstances. Sometimes we take this set for granted. We then assume that some set of *standard circumstances* (standard temperature, standard background noise, a standard set of equipment, or the like) is the case.

The idea of a standard circumstance is important and has to be developed in fact for every type of action. Some actions presuppose the existence of certain tools, some presuppose the existence of particular backgrounds. Other actions in fact also presuppose that other agents perform actions. (Compare the example of the minister writing a bill discussed above, pp. 69-70) Ability to do something in this case then entails *the ability of having things done*. This has interesting consequences for the interpretation of some examples in medical ethics. I shall return to this in the next section.

The physical and mental conditions of ability in relation to physical action have been extensively investigated above. But what about ability in relation to decision? Does the act of deciding require any ability? I think the answer is, yes. A person can be genuinely unable to decide about a matter. This can be due to weakness of will, to low intelligence but also to lack of knowledge of the matter to be decided about. It is also a logical point that one cannot decide to perform a particular action unless one at least *believes* that one has the power to perform (or the practical possibility of performing) the action in question.

Let us return now to the theoretical concept of autonomy or power to govern oneself. I have suggested that at least part of the analysis of this concept is tantamount to the analysis of the concept of practical possibility. And since an explication of practical possibility presupposes the specification of a set of actions — namely the set of actions that *A* has the practical possibility of performing — this also holds for autonomy. One can only be said to be autonomous in relation to a certain set of actions. The full analysis of autonomy, then, requires this specification. In particular we need to know which set of actions is involved in a person's governing herself.

Let me now attempt to formulate the extended concept of autonomy that comes out of this analysis:

> *A* is autonomous (in the sense of having the practical possibility) with respect to action *F*, only if *A* has both the ability and the opportunity to decide to perform *F*, and has both the ability and the opportunity to execute this decision, given a specified set of circumstances.

This definition requires supplementation of at least the following kind. Two very important additions are called for because of the possibility of an action's being *coerced or forced*. There are at least two kinds of force to be distinguished between. For the sake of simplicity let me call them *physical force* and *mental force*. For physical force I take the following paradigm: *A* forces *B* to move from one place to another by actually dragging *B* to the latter place. In this case it can first be observed that *B* does not perform any action of her own at all. But more importantly, *B* does not have the practical possibility of *not* moving from the first to the second place. And the latter element is central to our ordinary notion of autonomy. In order to be autonomous with regard to action *F* one must have the practical possibility of deciding not to perform *F* and also the practical possibility of not performing *F*.

But there is also another and more important kind of forcing an agent which does not exclude the agent's practical possibility of performing *F*, nor her practical possibility of not performing *F*. This kind of force has been analysed in some detail in chapters 4 and 6. It is also quite relevant for our ordinary notion of autonomy.

We considered above as a paradigm the following case. A cashier was at gunpoint forced to hand over some money to a robber. We normally say that the cashier is not autonomous with respect to this action, although she indeed has the practical possibility both of handing over and refusing to hand over the money. As we know, some brave cashiers indeed defy the robbers and refuse to give out the money.

The sense of autonomy that we wish to characterise in our ethical context, however, excludes also this weaker kind of force. There are therefore particular restrictions on the causal stories behind the person's deciding or acting in order for her to be an autonomous agent in a particular respect. The causal story should exclude all kinds of threat as characterised here. It should also exclude certain other kinds of external influence, such as deceitful persuasion and manipulation.[50]

ON THE NOTION OF A RIGHT TO AUTONOMY

So far I have presented a series of senses of a theoretical notion of autonomy, mainly including a person's practical possibility of action but also including a similar practical possibility of omission as well as a certain freedom in arriving at this practical possibility. The important question now is: what of all this do people have a right to have as well as execute? As a corollary of this question we may ask: what of all this should we respect? And what does such a respect amount to?

The answer to these questions would require a treatise of its own. Let me here only confine myself to giving a paradigm example and to making a few theoretical observations.

> *Paradigm right to autonomy in a medical ethics context*: A person has a right to have the power to decide and act on matters concerning herself and also the right to execute these decisions and acts unless they infringe on other people's parallel rights.

(Observe here that I use the word "power" for the complex notion involving both practical possibility and the right causal story behind the practical possibility.)

I think that this is a central kind of right in the field of autonomy. It is, I believe, central in health care and health promotion.

It is easy to exemplify how this complex right can be respected. In a medical ethics context a doctor or a nurse can respect this right by carefully informing a person about

---

50 These concepts are not further analysed here. For important treatments of the factors which compromise autonomy, see G. Dworkin [26], and R.R. Faden and T.L. Beauchamp [29], pp. 337-373.

possible alternative measures with regard to her health. The doctor or nurse can also give the person a fair opportunity (in terms of time and peace and quiet) to reflect on these alternatives before deciding. They can, finally, let the person speak out and say what she wants the treatment to be.

I think this is a standard description of the paradigm case of respecting autonomy in a medical ethics context, including the ethics of health promotion. The individual shall be able to arrive at a decision through a proper causal sequence involving adequate knowledge. The individual must not be forced to form this decision by any threat. But let me now analyse this case further in the light of my theoretical characterisation of autonomy.

According to the theoretical analysis, autonomy entailed the practical possibility of deciding and acting with respect to a certain action. To what extent does the standard procedure of paying respect really empower the patient to decide and act?

Consider first the decision part. Does the patient get the proper opportunity to decide? Yes, she is not disturbed and gets the reasonable time to ponder the alternatives. Is the patient really enabled to decide, in the strict ability sense? The procedure of information is supposed to secure that. But here we can ask our first critical question. The act of giving information is not tantamount to creating a state of insight within the patient; insight requires a proper apprehension and a minimal degree of intelligence on the part of the patient. Is it normally ascertained whether such insight has been achieved, or is this even normally required in ethical textbooks or codes?

So much for decision. But what of the act decided about? Here we immediately encounter an interesting problem which is very typical for the medical ethics context. The typical decision in this context is a decision about *permitting someone else to do something with regard to oneself.* The patient permits the doctor, or the patient permits some other carer, to perform a treatment or to take a health-promotive measure. The subsequent act is then performed by some other person than the deciding agent.

How, then, should the patient's right to autonomy be interpreted in this case? If the patient cannot perform the treatment on her own, then it is improper to say that the patient has autonomy with respect to this action. But can we then sensibly adhere to the patient's right to autonomy with respect to treatment? That would entail really enabling the patient to perform the treatment on her own. Or should the situation be analysed in some completely different way? Is it some other right that we are respecting?

Another way of interpreting the situation is to say — and indeed my theoretical description of autonomy allows for that — that the patient in fact has the power to be treated. As I said, to be able to do something may entail one's ability to have things done. The patient is then, according to this interpretation, autonomous with respect to the treatment. The doctor can respect the patient's right to autonomy by actually fulfilling her part in the treatment. (The situation is then theoretically similar to the undersecretary's writing a bill given the minister's decision.)

It seems to me that we need to become clear here about how to look upon this matter, not least when we go on to talk about the creation of autonomy. In, for instance, David Seedhouse's writings [88], [89] the idea of creating autonomy is central. What powers, more exactly, are we to create? Are we referring to the same powers as the ones that are central in the context of respecting rights? Is our main task to enable people to make informed decisions about themselves? Or is our task more generally to enlarge people's practical possibility of acting in various ways, including providing opportunities for their action? And does the empowerment also encompass people's power to make other people do things for them?

I think that my analysis has shown that there is a great complexity in the concept of autonomy, both in terms of type of power involved and in terms of the set of actions that the power is related to. An important area of deep research is open for exploration.

CHAPTER THIRTEEN

A SKETCH FOR A THEORY OF HEALTH ENHANCEMENT

GENERAL REMARKS ON HEALTH ENHANCEMENT AND HEALTH PROMOTION

In this chapter I shall propose a characterisation of the notion of health promotion, substantially using the model for analysing interaction presented above. Let me first stress a terminological point. The basic analysis that I propose is applicable to all kinds of interaction between people. Thus it is applicable to all kinds of interaction which could have the enhancement of health as a result. In previous presentations I have used the term "health promotion" as an abstract term for all such kinds of interaction. This use is contrary to some conventions in Anglo-American discourse on the matter. It is particularly unusual to include the whole field of medical (curative) care under the term "health promotion". Since my own focus in this volume is very broad I wish to keep the general connotation here. Hence the choice of the term "health enhancement".[51]

In this chapter I shall however put one restriction on the set of health-enhancing actions discussed. I shall confine myself to such interactions where the primary agent has the *intention* (although not necessarily explicit) to improve the health of some secondary agent. My focus will thus be primarily on the planning perspective. I am less interested in such actions as just happen to have healthy consequences. By a health-enhancing action in this context I thus mean the following:

> A health-enhancing action of *A*'s is an action by which *A* has the intention to maintain or improve the health of some individual *B*, where *A* and *B* can, but need not, be identical persons.

Observe, however, that this characterisation not only restricts but in a way also widens the set of health-enhancing actions. It includes also *unsuccessful* actions which are performed with the intention to enhance some person's health.

Health-enhancing measures can be of several types with regard to the agent of the action. I shall here distinguish between *individual, institutional and collective actions.* An individual health-promotive action is one with only one agent. An example of this is the mother who sees to it that her son puts on a raincoat on a rainy autumn morning; another example is the teacher who advises his pupils not to smoke. By an institutional action I mean, for instance, a piece of legislation that is put into practice by an institution in its name, for instance a parliament or a health education board. The main characteristic of an institutional action is that no particular person is responsible for the action (an exception being perhaps the case where the institution only consists of one individual). By a collective action I mean such an action as presupposes at least two

51 See my [67], pp. 11-12. R.S. Downie *et al.* [24], for instance, use the term "health promotion" in a more restricted fashion.

cooperative agents and where at least one of the participants has a health-enhancing purpose with her action.

Health-enhancing actions can be either more or less direct. By direct I mean an action which involves some direct manipulation of the subject's body. It is important to realise that such actions can be undertaken outside purely medical care. I have in mind such things as vaccination and preventive surgical operations. By direct manipulation I also mean the patient's own activities with body and mind, including abstention from certain activities. The action of building a sports centre, however, can only contribute *indirectly* to health-enhancing effects. The sports centre plays the role of providing the opportunity for other actions of a sports character, which in turn can affect the health of the people involved. The original action of building the centre, however, fulfils the condition of being a health-enhancing action in my sense if it is performed with the intention of improving the health of the population.

A particularly important type of indirect action comprises those activities which entail *interaction*. This is the case where a primary agent *A* intends to influence another agent *B* to act in a certain direction in order for the latter to maintain or improve her level of health.

A further classification of health enhancement has to do with variations in initial state and end state. The initial state of health of the secondary agent can vary from extreme ill health to extreme good health. The intended result of the measure in its turn can vary from the prevention of a very low degree of health from becoming worse, to the achievement of a state of complete health. The worst initial states are of course more common in health care proper; the maintenance and improvement of positive states of health are the objects also of agents and institutions outside health care.[52]

ON VARIOUS FORMS OF HEALTH ENHANCEMENT. THE INTERACTIVE CASE

Most forms of health enhancement — those which are generally discussed in the literature and which have the greatest ethical interest — are the indirect forms. Their influence on a particular person is then mediated by some further causal link, either some action or some series of natural episodes. Let me here distinguish between two main types of indirect health enhancement.

(a) *Health enhancement through change of the environment.* Here the logic of action is the following: *A* influences the health of *B* by changing some part of *B*'s physical or cultural environment. (The influence can here be even more indirect. *A* can try to influence someone else, for instance an authority, to decide to make a change in *B*'s environment. This case then also involves the subsequent type of health enhancement.)

52 A taxonomy for differentiating between health-enhancing actions along this dimension has been proposed by P-E. Liss [48], pp. 63-83.

(b) *Health enhancement through interaction.* By interactive health enhancement I mean the case where an agent *A* influences another agent *B* to perform a directly health-enhancing action.

In analysing the latter case I shall make extensive use of the schema for intentional explanation introduced in chapter 4. As in the general case of human interaction, one can influence the course of a health-promotive action by manipulating one or more of the components of the intentional explanans, and one can manipulate various combinations of them. Let me illustrate with a simple example of health education.

Let us imagine that a Health Education Board carries out research into the existence of mould in a block of flats. It appears that the situation is serious. Most of the investigated flats are affected. The risk of contracting asthma allergy is very high, in particular for the children in the affected population. As a result the Board informs the inhabitants of the block about the situation. The circumstances are described as highly dangerous to their health. Let us now assume that most of the inhabitants already possess a determinant of action corresponding to the following:

A has the standing intention of maintaining her own state of health

Through the information from the Board a possible component 2 is created or changed in *A*. *A* has come to know about the dangerous environment. As a result *A* can perhaps on her own form a component 3, i.e. decide what she must do in order to avoid the threat. Alternatively, she can be helped in this respect by the Board again. They can give her advice about possible measures she might take. As a result we can have the following explanans represented among most of the inhabitants of the block.

1. *A* has the intention of maintaining her health.
2. *A* judges that the mould situation in her flat is dangerous to her health.
3. *A* judges that she must carry out a complete sanitation of the flat in order to maintain her and her family's health.
4. *A* has the financial resources and has every opportunity to carry out this sanitation.

As a result of all this *A* performs the required action. This is an analysis of a health-promotive action of the information kind. Now, similar presentations can be made of other kinds of interaction, such as recommendation, advice, threat and persuasion.

ON INTERACTIVE MODES IN MEDICAL CARE

Many forms of medical care (within hospitals or primary care centres) involve the direct manipulation of a patient's body or mind, and these actions are thus direct forms of health enhancement in my sense. However, practically all forms of medical care presuppose other forms of interaction between carer and patient. It is hardly ever the case that a medical practitioner approaches a patient and without any question applies surgery or some other treatment to her. (The only such cases that there should be — and which can pass without violation of elementary ethical rules — are emergency treatments of unconscious, incompetent or insane people.) Prior to direct manipulation of a body or mind there is in the standard case of medical care a long session of communication between a patient and a carer.

This communication consists partly in the exchange of information, partly in questions and answers, partly in advice and recommendations.

A sketch of a schema for such interchange is the following:

(i) The patient presents a medical problem, typically in terms of bodily or mental signs and symptoms, which worries her or which directly reduces her health.
(ii) The carer tries to classify the problem or its possible internal causes in terms of the traditional medical classification, e.g. ICD [40] or ICIDH [41].
(iii) Through a series of questions and answers and through other modes of investigation, the carer, in the ideal case, reaches a diagnosis; the problem is classified.
(iv) The medical inventory is consulted to find an adequate solution to the problem. The result of such a search may, in general terms, be one of the following three:

(i) No action whatsoever can be recommended.
(ii) Some non-medical action can be recommended.
(iii) Some medical action can be recommended.

In the latter case the proposed medical action is presented to the patient. The patient is informed about its possible and probable consequences, and as a result agrees or refuses to take part in the recommended action. In some cases there is no single recommended action but a set of alternatives proposed. Among these the patient may have to make a choice or, again, may refuse to take part altogether.

It is easy to see how, in this series of events, many of the various modes of interaction, as classified above, come into play. The mode of *information* plays a prominent role, first when the patient presents her problem. But it also plays a role when the carer informs the patient about the nature of the problem as classified. In the ideal case the carer then proposes a course of remedial action (or a set of alternative actions). This presentation normally has the character of *advice* or *recommendation* (given the plausible hypothesis that the patient wishes to recover). If necessary, the carer also helps the patient by giving her the *opportunity* to take part in the mode of medical care suggested.

A plausible reconstruction of this central stage of interaction between the carer and the patient is the following:

(i) *A* desires to get rid of her medical problem.
(ii) *A* is informed by a carer that her problem is of type *T*, and *A* believes that she is in general situation *C*.
(iii) *A* is advised by the carer that the only way to get rid of her *T*-problem in *C* is to take part in medical action *M*.
(iv) *A* is given the opportunity to take part in *M*.

Observe that this explanans is not yet completely explanatory. Premise (i) contains the word "desires" and not "intends". Medical information and advice is normally given (or should be given) in a situation where there is still an element of choice. I have indicated in the syllogism that the patient has not necessarily decided yet. Before she actually *decides* to get rid of her problem she may wish to get detailed information about the medical action *M* and its possible side-effects. If *M* appears to be unproblematic or at least much less painful or worrying than the initial problem, we can expect her want to issue in an intention to remove the problem.

THE IDEA OF HEALTH EDUCATION AND SOME INTERACTIVE MODES CONNECTED WITH IT

The field of health education is logically much more complex than that of the standard forms of medical care. In the latter case most series of interactions start with the presentation of a patient's problem. This immediately gives the subsequent process a clear focus. Health education very often does not have such a focus. Indeed, health education has the same conceptual generality as health enhancement itself (although there are instances of health enhancement which are not instances of health education).

First, health education is general in that its target can be the health of the whole of mankind and not only the health of a single individual or of a group of individuals. Second, health education need not be directed towards preventing a specific disease, such as lung cancer, in the population. Third, health education is *not* restricted to the area of medical care; it is not solely concerned with illness or the danger of illness. Fourth, health education involves the delivery of information (as well as other strategies) concerning the whole area of health enhancement, involving all the concepts and causal sequences which the present chapter is intended to elucidate. Hence, health education must conceptually grasp the whole area of health enhancement.

Having offered these general statements let me attempt to give a preliminary characterisation of health education:

> Health education is an activity directed towards a defined population and intended to enhance the health of this population by increasing their theoretical and practical knowledge and/or by changing their attitudes towards health and health behaviour.

Observe that (as in the general case of health enhancement) I have not conceptually required that a particular health education strategy must be successful. Whether it is successful or not will be a matter for empirical discovery.

Let me now analyse what the success of a health-educative strategy can amount to and how it can be characterised. Consider the following concrete situation. A Scottish educationist launches a particular anti-smoking programme. He aims at reducing the cigarette consumption in the city of Edinburgh by 50% in a year. In order to realise the goal of this programme the educationist uses the press, radio and TV, as well as a great number of public meetings. Through these means a lot of biochemical and medical information is presented. The public are informed about the chemical effects of smoking on the body and about long-term consequences of smoking with regard to bodily and mental functioning, both general function-reducing consequences and particular effects in terms of severe diseases and impairments.

The information is linked to recommendations (as well as persuasion and training) with respect to alternative activities and with respect to measures to be taken to get rid of the smoking habit. Medical and general advice is given and concrete help is offered to those who might have difficulty in becoming free from their addiction to smoking.

The final success of this programme — the reduction of cigarette consumption in Edinburgh by 50% — is already defined and obvious. But it can be analysed into a number of steps, which can be characterised in my theoretical terms.

First, there are the four steps involved in creating a relevant intentional explanans with regard to one particular individual. Each individual is *informed* about the habit of smoking and its probable consequences. Each individual is informed about the necessary means for avoiding these negative consequences, viz. refusing to start smoking or trying to stop smoking with the help of various facilitating devices. Moreover, each individual is offered individual advice and help to avoid the addiction. We may also assume that certain municipal authorities are supporting the project by prohibiting smoking at places of work and official buildings, thereby creating greater opportunities for success.

Health education then obviously involves the influence of components 2-4 in the intentional explanans with regard to each target-individual. The programme has succeeded to some extent, taken one step forward, for each belief or each opportunity in the right direction that has been realised.

But does health education have anything to do with the first component, the intention-clause? Does health education (and should it) aim at changing people's goals and attitudes? This is an urgent question, since recent research into health behaviour has stressed that the key factor with regard to change in behaviour is a transformation of attitudes, and that these attitudes are intimately connected to desirable life-styles. A change in life-style is not accomplished, it is claimed, without attitudinal change.

A full discussion of this issue, with its ethical complications, must wait. Let me here only stress some logical points. From the perspective of our theory there is an obvious, not only empirical but conceptual, truth in the dictum about attitudinal change. Every intentional action presupposes a complete explanans, i.e. the adequate intentional clause. In order for the background information (about the situation and necessary means to the healthy end) to be operative the recipient must have formed an adequate intention. In the straightforward case she has explicitly formed the intention to raise or maintain her own level of health or she has been persuaded to form this intention.

Another, more interesting and controversial, alternative is that she has formed the intention to attain some *other* goal, for which the necessary and sufficient means as a matter of fact are also sufficient for the promotion of her health. Assume that the health educationists in their advertisements were to use some popular sportsmen or pop singers as ideals. The kind of attitude they aim at creating (and would be likely to create) in young men or women is one of wanting to imitate the idols. If this results in an intention it is an intention to imitate the idol and not primarily to promote health. The net result, however, given that the educationists' background analysis is correct, is some degree of health enhancement.

The logic of the attitudinal change needed has already been analysed. Component (1) in an explanans is created by showing that the goal entailed is conducive to or even necessary for the attainment of a higher-order goal of the agent in question. This again means essentially the provision of information about means to ends. An advertisement indicates, for instance, that living the life X lives is the most expedient way to hit it off with members of the opposite sex, which to the agent is a highly desirable end.

So much for the first major step in the chain of succeeding with an act of health education. We can be very brief concerning stage 2. This is the stage which is often called *compliance*, viz. the agent does what she is recommended to do. The reason why we can express ourselves in a laconic way is the following. Given that a complete practical syllogism is formed, then the compliant action must follow with logical necessity. Given a complete step 1, then step 2 must necessarily follow.

To state this feature of the theory of the practical syllogism is not to retreat to the ivory tower of the philosopher. It is just to say that the various kinds of imperfections of reality — the facts that can prevent compliance — are, strictly speaking, to be treated within step 1. Consider the following list of things that may go wrong from the point of view of the health educationist:

(a) The agent (*A*) has not *really* changed her attitude; she does not really intend to improve her health; she may want it in a loose sense, but there are so many other things that she wants more, e.g. indulging in night-life.
(b) *A* has not really got the message; she is not convinced that smoking is dangerous to her health, at least she does not believe that it will affect her body in the foreseeable future.
(c) *A* is not convinced that she must change life-style in order to avoid the dangers of smoking, or *A* has not been adequately informed about what she needs to do to avoid the dangers.
(d) *A* has not been given any real help to change her habit. She is an addict, and quite substantial measures must be taken to enable her to become free of her addiction.

These and other, similar variants are the typical preventive factors. They all involve various types of logical defects in the practical syllogism leading to the desirable course of action. The work of eliminating all such preventive factors can then be repeated for the number of individuals needed to account for 50% of the cigarette consumption in Edinburgh.

So much for the success of a particular health-educative programme. But observe that, strictly speaking, this is not enough for the success of an act of health promotion. A reduced consumption of cigarettes is not the same as improved health. The reduced consumption is only a very plausible cause of a sustained and improved level of health in the population. The causal mechanisms can be various. Observe first that there are two target-populations, not only the habitual smokers but also the *potential beginners*, i.e. those who might have started smoking during the period of health education. For the latter group the campaign functions as an act of health support. A potential danger to health is eliminated and thereby the health of the target-population is prevented from being reduced.

The health of the habitual smoker can be affected in at least three ways. First, certain pathogenetic processes, which are due to the smoking, can be aborted. A growing incidence of pneumonia or lung cancer can be prevented. Thus again there is a case of health support. Second, a disease which has already broken out can be halted and even reversed when the negative influence of smoking is no longer there. Third, the general function of the person's body and mind can be immediately raised as a consequence of her becoming free of her smoking habit. The latter two cases are instances of health improvement.

CHAPTER FOURTEEN

ON THE ROLE OF COMPULSION IN MENTAL ILLNESS. THE CASE OF FORENSIC PSYCHIATRY

INTRODUCTION

I shall now consider a further area of application for some of my action-theoretic ideas. This is the area of interface between pscyhiatry and jurisdiction. My discussion concerns basic issues in legal theory. In focus is the question of the criteria for a person's legal responsibility for her deeds. Such issues are continually discussed in international legal theory. (For a recent account, see Ten [100].) In this chapter I shall allow myself to approach the area of criminal responsibility by referring to a special case, viz. the development of Swedish criminal law and Swedish forensic psychiatry during this century. This specification will not prevent the general theoretical assumptions from standing out.[53]

THE PROBLEM OF ACCOUNTABILITY IN SWEDISH CRIMINAL LAW

In 1965 the hundred-year-old Swedish Criminal Code was replaced by the new Penal Code [61]. This was an important event in many respects. It implied not only a general modernisation of the penal statutes, it meant also the formal acceptance of a new philosophy of law which, compared with the thinking of the previous century, was revolutionary.

The new philosophy was the positive (reformist) philosophy, which claimed that the fundamental aim of punishment is crime prevention and that such prevention is most effectively realised through the reform of individual criminals. The Penal Code thus explicitly repudiates the old retributivist ideas about punishment. According to these the society has a right (or even a duty) to inflict suffering on a condemned criminal in proportion to the severity of the crime. This right should be executed regardless of the possible consequences of the punishment.

The reform had important implications also for forensic psychiatry. In the old days such criminals as were insane or gravely mentally deficient had been treated in accordance with the traditional theory of accountability. (For a discussion of this concept see below in this chapter.) A criminal who was judged to have been insane at the time of his deed was *ipso facto* considered non-accountable. Therefore he was not responsible for the crime and could not be punished. An insane criminal was often taken into psychiatric care and ended up in a mental hospital, but this was not as a result of a criminal

53 The presentation here is a summary of some of my main ideas in [68].

sentence but instead as a result of a medical decision according to the medical legislation.

The reformist positive philosophy of law does not recognise the concept of accountability. Everybody, be he mentally healthy or insane, is responsible for his deeds. Therefore society should react equally much to destructive actions of the same type regardless of the mental health or otherwise of the perpetrators. Today, therefore, the insane criminal is judged to be responsible for his deed and can therefore be *sentenced* by a judge to the sanction of psychiatric care. This is a change of great philosophical interest and also — I believe — of some practical importance.

THE NOTION OF ACCOUNTABILITY IN CRIMINAL AND CIVIL LAW

I shall not here try to finally settle the conflict between the old retributivist and the new reformist philosophy of law. Let me here only see how much an action-theoretic analysis can do in order to clarify and perhaps dissolve some elements of the conflict. A starting point for this analysis is to look into the notion of accountability, which was a central notion in the old philosophy and which the new philosophy purportedly has expelled from the arena.

What was accountability? Consider here the classical list of criteria that has dominated legal thinking in Europe during at least the last two hundred years. A perpetrator — since the vast majority of perpetrators are male I shall in the following use the male pronoun — can, according to this thinking, be considered accountable or responsible for his criminal deed, if and only if

(i) he knew what he did when he did it
(ii) he knew that what he did was wrong
(iii) he had a free choice, i.e. he could act otherwise than he actually did.

(Clause (ii) is here simplified. In most legal systems sheer ignorance of the criminal nature of an action is not sufficient for a person to be exempted from penal responsibility. An offender has normally been regarded as non-accountable in this respect *only if he lacked the capacity or opportunity to know* that what he was doing constituted a breach of a moral or legal norm.)

I shall here also add a further requirement which is always held implicit, but which because it is only implicit can often be forgotten in the theoretical discussion. This requirement is the most basic one.

(iv) A perpetrator is accountable for the criminal deed *D* only if he has actually committed *D*.

This clause may seem trivial and self-evident. It may even be viewed as a simple logical consequence of condition (i). A person who knows what he is doing, it may be argued, must also intentionally perform this doing. The trouble with this answer, however, is, as

I shall show in detail below, that neither his knowledge nor his intention need concern the deed *under the description under which it is a criminal deed.*

In my general introduction to the concept of action above I noticed among other things the great complexity of many action types. Many actions require for their existence a complicated set of conditions. It is then not sufficient that a certain behaviour or that a certain set of bodily movements has been performed. Other conditions must also be fulfilled. I gave the example of getting married. For a marriage to come about is required — in addition to certain bodily movements — at least the following: existence of a partner, publishing of the banns, consideration of impediments, person to perform the ceremony, witnesses and authorised premises.

The situation is similar with crimes. Crimes presuppose for their existence a number of prerequisites, often divided into objective and subjective. Among the objective prerequisites are the physical movements (in some cases even omissions) but also certain results as well as circumstances. But in addition to these there are also normally a number of so-called subjective prerequisites. Most of these prerequisites have to do with the intentionality condition introduced in chapter 1. I said there that for a bodily movement to count as an action it must be intentional, i.e. be preceded by an intention to realise a certain state of affairs.

The simplest and most clearcut legal case is that of *direct intent.* ("Intent" is the legal term for intention.) By this is meant the case where there is congruence between intention and action in the following way: the intention covers exactly the action which is regarded as thereafter having been performed. Take for instance the case of a person who intends to burgle a flat and steal a TV-set. As a result of this intention the person breaks into a flat and removes a TV-set from the flat. By the combination of the intention and the objective happening the person has fulfilled the prerequisites of the crimes of burglary and theft of a TV-set.

Lawyers distinguish also other species of intentionality. To designate the following typical case is used the expression *indirect intent.* A perpetrator has the primary intention of bringing about the effect *P*. He does not really intend to bring about *Q*, but is at the same time aware that *Q* is a necessary condition (or consequence) of the realisation of *P*. From the juridical point of view he has thereby had the indirect intent of bringing about *P*. Consider the following example:

> The perpetrator blows up a boat with people on board, and he does so in order to get the insurance money for the boat. The purpose here is financial gain, and the death of the people on board is merely a necessary but not in itself desired side-effect.[54]

For certain crimes to be regarded as having been committed there is however not required any of the mentioned forms of intent. Such crimes can be brought together under the heading of negligence (*culpa*). A typical example is the following. A boy is kicking a football about. He miskicks, and the ball shatters a window of the house next door, injuring the neighbour who happens to be looking out. Here the boy does not even have

54 A. Nelson [62], p. 37, my translation.

the potential intent of damaging the neighbour's window. But the act is characterised by negligence: the boy ought to have thought, and might have realised, that his kicking of the football might have the consequences it did have. He is therefore to be held responsible for the damage and injury that resulted.

Thus the nature and scope of the intent may vary. But it is important to note (and this is indeed also entailed by my notion of action) that all the cases considered (including the *culpa* case) involve in a basic sense intentional actions. What is problematic, and requires the drawing of subtle distinctions, is the matter of under what description the actions are intentional ones. Even in the case of the boy who by shattering the window-pane injures a neighbour, there is at bottom an action that has been performed with a direct intention, namely the kicking of a football. It is true of all my examples that they involve more than bodily movements (more than reflexes or the twitching of muscles): they involve action that contains a degree of intentionality.

But my analysis of crime is not yet complete. There are actions which do evince the above-mentioned objective and subjective prerequisites but which are nevertheless not finally classified as crimes, and the reason is that certain disqualifying additional conditions are present. Thus a person who has wilfully performed a punishable action can be exempted from penal sanction if certain exculpatory or justifying circumstances are present. Foremost among the latter is the case of *self-defence*. The first sentence of Chapter 24, §1 of the Swedish Penal Code reads as follows:

> An act committed by a person in self-defence is not punishable.

Exculpatory circumstances involve various forms of compulsion. One such variant is the compulsion of duty. When a policeman or member of the armed forces wilfully performs a punishable act, but performs it in order to comply with regulations or on direct orders from a superior, his act shall in principle not be punished. Another form is sheer physical compulsion, when, for instance, a person is forced at gunpoint to commit a crime.

Given the observations so far I can construct a concept of crime — still from a legal point of view simplified — which has the following complexity:

A person *A*, who has the reached the age of criminal responsibility, has committed a crime *C*, if and only if,

(i) *C* is punishable according to the Penal Code
(ii) *A* has caused the behaviour which constitutes the behavioural aspect of *C*
(iii) *A* has caused the behaviour wilfully or by negligence
(iv) other objective prerequisites for *C* (circumstances and results) have been realised
(v) neither *A*'s intent (or negligence) nor his behaviour or its result was exculpated or legitimated by self-defence, compulsion or the requirements of duty.

OLOF KINBERG'S CRITICISM OF THE CLASSICAL THEORY OF ACCOUNTABILITY

As I have indicated in my introduction, the classical idea of accountability has, formally, vanished from the Swedish penal law. This is mainly an effect of the legal and philosophical analysis performed by the forensic psychiatrist Olof Kinberg and the professor of criminal law Karl Schlyter. Their contention was that accountability in its classical interpretation was an *untenable* concept. Kinberg, who did most of the philosophical groundwork, directed his main attacks against the third criterion for the application of the concept, viz. the criterion that the perpetrator must have been able to perform some other action than the one that he in fact performed. This criterion, according to Kinberg, is never fulfilled. No person can ever perform something else than what he actually performs. Hence nobody can ever be accountable. Hence further, the criterion cannot be used to distinguish, in a useful way, between such people as should be punished for crimes and such as should not.[55]

Let me put the reasoning in more philosophical terms. According to Kinberg the classical concept of accountability presupposes that some version of *indeterminism* is true. The contradictory thesis, *determinism*, contends, in its ontological version, that all events in the physical world are sufficiently conditioned by a conglomerate of previous events. Assume that the set of events *E* is such a sufficient condition of the set of events *E1*. Assume also that we have a complete description of *E*, and know the universal law which connects *E*, under this description, with *E1*, under a certain description. From this follows an *epistemic* version of determinism, which in general says: Given that we have complete knowledge of a certain state of the world, and knowledge of relevant universal laws, we can in principle predict the next state of the world.

If determinism is true then all events, including human actions, are sufficiently determined by some previous events, perhaps in the case of human actions by a set of neurological events. If every action is determined, then in one sense of the word every action *must* be performed. But if every action that is in fact performed also must be performed, then no agent can ever do anything else than what he actually does. Hence, if criterion (iii) of the accountability concept is to be meaningful complete determinism must be false. Therefore this criterion must presuppose some version of indeterminism, entailing at least that there are some events which are not sufficiently determined. The paradigm example of such an event is the action performed by a sane human being. The protagonist of the accountability concept cannot, on the other hand, hold the most extreme version of indeterminism, which entails that no event whatsoever is sufficiently determined. The reason for this is that some events, in particular the actions of the mentally insane, must be considered to be sufficiently conditioned.

The protagonist of the accountability thesis must then, according to Kinberg, hold that some weaker kind of indeterminism is true; a kind of indeterminism which allows that some human actions issue from a free (uncaused) choice. This is, however, according to Kinberg, a metaphysical thesis which is in conflict with modern science. Let me here quote from Kinberg [45]:

---

55 For a modern exposition of a reformist philosophy of law, see K. Menninger [58].

> The central principle of classicism, imputability as dependent on freedom of will — which imputability sometimes exists and sometimes does not exist in respect of the same individual, and which is at the same time freedom (when it comes to choosing between different possibilities of action) and conformity to law (through physical influence in the shape of various types of illness and bodily deficiency) — has in the encounter with the consistent determinism of positivism had its position further weakened.[56]

Kinberg says that if the determinist doctrine is true, then no distinction can be made between persons who in the classical sense are accountable and those who are not. Since modern science, he says, presupposes the determinist postulate, it is not compatible with a modern scientific point of view to speak in terms of accountability or non-accountability.

Is Kinberg right, then, in claiming that modern science presupposes the determinist postulate? And does indeed the classical idea of accountability have anything to do with the question of determinism/indeterminism? I can be very brief with regard to the first question, for the simple reason that I intend to answer the second one with a no. The crucial expression "could not have acted otherwise" does not have to presuppose an indeterminist point of view in order to be meaningful. As a matter of fact it can very reasonably be interpreted within a completely different theoretical framework which is perfectly compatible with both a determinist and an indeterminist metaphysics.

Whatever the more precise definition of determinism, it is not for the theorist a thesis which can be proved, either by natural science or by any other science. Like the different postulates of indeterminism, it belongs to the class of metaphysical postulates that are prerequisites for various types of empirical research. The general determinist postulate has been a fruitful point of departure for a lot of scientific research, at least since the Newtonian scientific revolution of the 17th century. But with regard to modern microphysics it has been subject to considerable dispute: within the framework of the successful theory-making of quantum physics it is in the opinion of many not in principle possible to indicate a set of conditions which together are sufficient for certain microphysical events. Thus determinism does not here function as a universal postulate.

This short commentary is sufficient for my purposes here. None of my further argument is dependent on the determinist postulate. It may be easy to agree with Kinberg that the idea of a metaphysical free will, not determined in any way whatever, is difficult to grasp. But I do not regard this idea as necessary in order to defend a serviceable concept of accountability which can play roughly the same role as it now does in international criminal law (and as it did in Swedish criminal law up to 1965).

I shall defend the concept of accountability from two angles. Both have links to the preceding action-theoretic analysis.

1. The prerequisites for crime are in a sense at the same time prerequisites for legal capacity. Thus the concept of crime itself already presupposes a minimum concept of accountability.

---

56 Most of Kinberg's ideas presented here are to be found in [45]. The present quotation is taken from pp. 67-68, my translation.

2. The expression "could not have acted otherwise" can — and plausibly should — be linked with a concept of compulsion which is not identical to the concept of sufficient condition. Compulsion can be given a clear and distinctive definition in completely different terms.

ON THE PREREQUISITES FOR CRIME AND LEGAL CAPACITY

For it to be said that there is a crime, it is not sufficient that certain objective prerequisites be fulfilled, i.e. that a certain bodily movement be carried out and that certain physical consequences of this bodily movement be realised. Further required is the fulfilment of certain subjective prerequisites, first and foremost that there be an intention behind the act, presupposing in its turn (under normal circumstances) that the agent have a degree of understanding of what he is doing or has done. If I wilfully murder Smith, I am normally aware that I am doing so.

In the case of certain types of grave mental aberration there is neither adequate understanding nor intent. Thus if a person whose mental aberration is complete has without intent caused the death of another person, he has not committed the crime of murder or the crime of manslaughter. Such a person cannot therefore — and this is the position of the Swedish Penal Code — be held legally responsible for either of these crimes. With regard to most crimes the prerequisite of intent has the import that the perpetrator has to know what he is doing when he performs the punishable action. We have indeed already seen that the first condition in the classical theory of accountability is that the perpetrator have an understanding of what he is doing. It must therefore be justifiable to say that the Penal Code in the very concept of crime involves at least a *minimum* theory of accountability.

I have discovered no indication that any representative of the positive school has analysed crime from the point of view of action theory. Neither in the general determinist postulate nor in the idea of social responsibility is there drawn a clear distinction between pure bodily movements and such intentional actions as have the status of crimes. On the contrary, indeed — in the positive ideology there is emphasis on the legal and social similarity of these phenomena.

But for all criminal law, including the present Swedish, it is of great importance in what way a bodily movement or action is determined. Among the sufficent conditions for acts are intentions and beliefs, and the specific nature of these can be decisive with regard to the identification of actions as actions of a particular type. More especially: a crime can often not be identified as a crime of a certain type unless it has been preceded and determined by a particular intention. The crime of murder, for instance, requires for its existence that a person shall have performed a series of actions with the direct intent of killing another person.

Let me clarify the problem by way of a number of examples:

1. A gale causes a tree to fall and kill a person.
2. The gale knocks over Smith, who falls against Jones, who falls so unluckily that he dies.

3. Smith has an epileptic fit and falls against Jones, who falls so unluckily that he dies.
4. Smith cuts down a tree, without noticing Jones, and the tree falls on Jones and he dies.
5. Smith is forced, under threat of himself being shot, to shoot Jones, who dies.
6. Smith is paranoid and regards his life as being endangered by Jones, therefore he shoots Jones and Jones dies.
7. Smith shoots Jones under the effect of intense emotion, and Jones dies.
8. Over a long period Smith cold-bloodedly plans the murder of Jones, and when the right moment comes he shoots him, and Jones dies.

What all these have in common is that they involve a person's death. Each in its different way gives a causal explanation of this death.

All of the examples except one involve another person as part of the causal chain. Within the perspective of the determinist postulate it can be said that Jones's death is in every instance sufficiently conditioned.

Against the phenomena here referred to as causal agents (including the gale), society can and should intervene. It is a principle that a person's premature death is to be prevented. Possible risk factors are to be eliminated. Against a gale we can for instance protect ourselves by putting up strong fencing. Or injury resulting from an epileptic fit can be avoided by, say, keeping epileptics in a sheltered environment — but of course best by effectively treating the epilepsy. Injury caused by a paranoiac can be prevented in corresponding ways. Cases 5, 7 and 8 are cases for traditional police intervention.

If, though, the mere fact that X is part of the cause of Y is sufficient for maintaining that X is socially responsible for Y, then all of the causal agents here — including the gale — are socially responsible. Now neither Kinberg nor any other representative of the positive school generalises in this way, and indeed Kinberg explicitly says that social responsibility only means that if any person's action possesses significance from the point of view of society, then that person becomes the object of certain reactions on the part of society. But this limitation seems to be merely conventional. By tradition preventive measures directed against natural phenomena do not form a part of jurisprudence.

But so long as the positive school does not define the concept of action, there will be no fundamental difference between the cases 1-8 from the point of view of social responsibility. If Smith is a part of the sufficient condition for Jones's death, then Smith — according to the doctrine of the positive school — is to be considered socially responsible for Jones's death, and is to be subjected to a sanction by society.

As I have already indicated, this is alien to all jurisprudence, including the Swedish Penal Code. Smith becomes penally responsible only if he has committed a crime, and in several of the cases in which he is here involved he has in fact not committed a crime. Cases 2 and 3, for instance, are normally not crimes. Smith has not performed an intentional act — Jones dies as the result of a mishap. Despite the fact that Smith has been a link in the causal chain that leads to the mishap, he cannot in any juridical system be held responsible for what happened.

Case 4 involves a wilful action, namely that of cutting down the tree, but it is only under very special circumstances that Smith can be said to have committed a crime. It would have to be demonstrated that he has acted with great negligence and that he ought to have expected that someone might be standing in the way. In such a case he can be regarded as having committed the crime of manslaughter. Case 5 involves an action which is probably not a crime either, since there is compulsion. Only cases 6, 7 and 8 involve actions which clearly fulfil the prerequisites of crime and which can lead to penal sanction.

Let me sum up. Smith is in seven cases a part of the sufficient condition for Jones's death, but in only three of these cases has Smith committed a crime for which penal responsibility can be enforced. Thus the concept of crime in itself requires an advanced concept of legal capacity.

Against this background there does not appear to be such a great difference between the Swedish criminal law of the past and that of the present. The prerequisites for crime are to a large extent the same in both systems. But one difference is fundamental: according to the older law a criminal act was not to be influenced by mental disorder, whereas nowadays it can indeed be so influenced — though under the important proviso that the prerequisites are still fulfilled. Thus even now, if mental disorder prevents the fulfilment of a prerequisite, there is no crime and thereby no penal responsibility.

ON THE STIPULATION "COULD NOT HAVE ACTED OTHERWISE"

The really difficult concept when it comes to discussing the classical theory of accountability is that which is expressed by the words "could not have acted otherwise". Certain persons are regarded as not having been responsible for their acts because in the situation in which they found themselves they could not — in some sense — have acted otherwise than they in fact did.

Traditionally this has been interpreted in such a way as to mean quite simply that acts performed by those who are not responsible are empirically sufficiently determined, with the evident implication that acts performed by responsible persons are not thus determined. Thus in this interpretation the theory of accountability would presuppose that indeterminism may be valid. It also presupposes, however, that certain acts are fully determined, among others precisely those performed by persons who are not responsible.

Kinberg — as we have seen — reacted strongly against this doctrine. There is no reason for supposing, he says, that the actions of so-called responsible persons should not be completely determined; according to a scientific view of man it is indeed absurd to assume this partial indeterminism. His conclusion is therefore that all persons are equally responsible for all their actions, and this general form of responsibility he calls (as mentioned before) social responsibility.

Kinberg's determinist thesis may be right. I am in fact uncertain about it, but it can anyway be accepted for the sake of argument. But from this it does not necessarily follow that the concept of responsibility is without content. For the expression "could not have acted otherwise" can be interpreted in another and indeed more reasonable way. According to the analysis I propose, the expression "Johnny could not do anything other than *F*" is to mean that Johnny was forced to perform the action *F*, and *not* that

Johnny's performance of *F* was sufficiently determined. It is a complete mistake to identify compulsion with sufficiency of determination.

As a matter of fact it is easy to convince oneself that a coercive circumstance with regard to an act is not the same as a circumstance making the action causally necessary. Consider the following clear-cut example. A person *A* is forced under pistol-threat to rob a bank. The pistol-threat against *A* is the coercive circumstance, the robbery is the action. But from the pistol-threat alone it does not of course follow with causal necessity that *A* will rob the bank. Many different outcomes are here causally possible: *A* may faint, *A* may try to run away, *A* may quite simply accept the consequences of the pistol-threat and let himself be shot dead. To say this, however, is not to plead the case of indeterminism, but only to say that the sufficient condition which involves the pistol-threat to *A* must also contain other factors, first and foremost *A*'s judgement of the pistol-threat as a coercive circumstance.

In chapter 4 in this book I have attempted to analyse more closely under what conditions a circumstance is conceived of as coercive, and have elucidated this within the framework of an action-determinist model. The basic import of this analysis is that a circumstance is coercive for an agent if it seriously threatens the realisation of an intention which the agent is not prepared to abandon.

A pistol-threat is coercive for every person who has the intention of staying alive and who is furthermore not prepared to abandon this intention. To stay alive there is only one thing to do, namely to do what the person with the pistol says. Thus in order to stay alive, *A* could not have acted otherwise than he did.

A's act is thus a necessary condition for the realisation of his intention. But his act is not necessary in the sense of being causally necessitated by the coercive circumstance, as can readily be seen from the fact that the coercive nature of the pistol-threat is at once dissolved if *A* as a matter of fact abandons his intention and is prepared to let himself be killed.

Now of course intentions which we are not prepared to abandon can be of very different character and be very different from person to person. The intention of staying alive is unusually universal, as is that of protecting oneself and one's family. But there are also other intentions — e.g. maximising one's profit, acquiring a beautiful wife or making a career for oneself — to which some persons cling with the utmost tenacity. Here, too, there can be coercive circumstances. For instance Robinson is intent on constantly maximising his profit, therefore on 14 October 1987 he was forced to sell all his shares. Characteristic, though, is that when it comes to the less universal type of tenaciously held intention, the intention has to be spelt out if the coercive nature of the circumstance is to be comprehensible.

As noted before, certain situations involving compulsion are of significance with regard to criminal law. For instance Chapter 24, §1 of the Swedish Penal Code begins as follows:

> An act committed by a person in self-defence is not punishable.

Nor are certain actions punishable which are ordered by legitimate authorities.

Certain actions which are brought about by compulsion are thus not regarded as crimes, and thereby do not involve penal responsibility. This is of interest for several reasons. The Penal Code, which in principle considers all persons to be equally responsible for all their actions, here makes an explicit exception for acts that have been performed through compulsion. Clearly the reason for this special treatment cannot be that these actions are more determined than other ones, for the initial hypothesis was that all actions are equally determined. The reason is instead that certain types of compulsion make an action excusable or *justified*. The act is accordingly exempted from penal sanction.

But what, then, are the types of compulsion that make an action excusable or justified? It can scarcely, of course, be a question of compulsion in respect of any and every sort of tenaciously held intention, e.g. the intention of maximising one's profit. It has to do, instead, with compulsion in respect of a very limited class of intentions. Occupying a special position is the intention of saving one's own life and the lives of others, and of avoiding serious injury, which brings in self-defence, as referred to in the above-quoted section of the Penal Code. To the same special category belongs the intention of doing one's duty. These intentions are of such a nature that it is not simply the case that practically all individuals tenaciously hold them, but also that society supports them (or at least accepts them). That is to say that if in any situation the realisation of such an intention requires that the agent perform a punishable action, this action is excused and no crime is regarded as having been committed.

But what has this type of compulsion and exculpatory circumstance to do with mental illness? First and foremost, it opens our eyes to the fact that the expression "could not have acted otherwise" acquires an important new meaning — a meaning, I intend to argue, which it has when applied to certain mentally ill persons who commit crime. The logic of action in respect of many (indeed perhaps all) cases of so-called *inner compulsion* is the same as in the case of compulsion which depends on an external circumstance.

The parallel can most easily and most clearly be seen if we look at the case of the paranoiac who believes that his life is threatened.[57] Subjectively, the paranoiac is in the same situation as *A* above: he sees his life as being threatened, and sees only one way of parrying the threat. The intention is in both cases one which is legitimised by society. The conclusion which ought to be drawn is that the paranoiac is exculpated in just the same way as any other person who performs a punishable act under threat of death.

It may be objected, though, that there is an important and indeed fundamental difference: the threat experienced by the paranoiac is merely imagined, whereas that experienced by *A* is real. (Even if we say that the threat to *A* may not be real, we do say

57 There is an old case illustrating exactly this situation of compulsion. This case has given its name to a classical rule in British criminal law, called the *M'Naghten rule*. M'Naghten had killed the secretary of Robert Peel, the British Prime Minister, in 1843. It came out at the trial that it was actually the Prime Minister himself that M'Naghten had intended to kill, the motive being his deluded idea that he was persecuted by the Tories. He thought that the only way to put a stop to this persecution was to murder Peel. M'Naghten was acquitted on account of his mental aberration. For a detailed account of the M'Naghten rule and other rules in forensic psychiatry, see J. Radden [82].

nevertheless that *A's* belief in it is well-founded. In the case of the paranoiac, however, the belief is not well-founded.)

But is it reasonable to require that such a belief shall in the objective sense be well-founded in order for an action in accordance with it to be regarded as excused? Is it not sufficient to show that the paranoiac in all seriousness considers his belief to be well-founded, and that no rational factor can change it? Subjectively he acts just as much under threat of death as does the normal person. If we do not accept that the paranoiac's act performed under compulsion is excused and thereby not to be regarded as a crime, we find ourselves in the bizarre situation of asserting that the person whose punishable act is based on reason and able to withstand objective scrutiny can be exculpated, whereas the person whose rationality is limited cannot.

Is it possible, though, to generalise from the case of the paranoiac, to extend the paradigm to other cases of inner compulsion — for instance to the type of compulsion which is more in the nature of a drive? Parallels are often drawn in the psychiatric literature between compulsion neuroses like kleptomania and pyromania on the one hand, and basic physiological drives like hunger and sexuality on the other.

It may appear at first glance that there is a very great difference between the experience of a drive and the experience of a threat, but in fact I propose basically the same analysis now as before. In the first place it is evident that a drive, however "strong" it may be, is not a sufficient condition for the action of satisfying it — any more than the threat was a sufficient condition for the relevant action. To convince oneself of this one might study the behaviour of the kleptomaniac or pyromaniac who is under constant police surveillance. There is little doubt that one would find that quite a lot of the "coercive" impulses were being suppressed.

Every normal person knows that he can subdue such drives as hunger, thirst and sex — the manifest presence of these drives does not have to be followed by action to satisfy them. Thus a drive, like a coercive external circumstance, is just a part of the complex which constitutes a sufficient condition for an act.

How, then, is the logic of action to be reconstructed in a case involving a drive? I think that this, if one is to go into detail, is a more complicated affair than the analysis of the threat situation. It would seem likely, moreover, that drives differ from one another to a fair extent. But here at least I can draw attention to two phenomena that have a central position with regard to the determination of acts by drives. In the first place every unsatisfied drive is a cause of discomfort, often in the form of a tension. In the second place an agent often knows from experience what the satisfaction of the drive bears with it in the form of pleasurable sensation.

Most people have set intentions which prima facie motivate drive satisfaction. The most extreme type of such intention is that of maximising pleasure. Less extreme types concern the maintenance of a physical state of relative peace and harmony. When it comes to the basic drives of hunger, thirst and sex, there arise of course also intentions which affect the continued existence of one's own person and of the human race. Thus the sensation caused by the unsatisfied drive leads a person to realise a basic intention, which can have a somewhat different character from person to person.

Now, are these sensations coercive? Certainly they can be so in the neutral sense of the word, where the perpetrator is not prepared to abandon or modify his intention (e.g.

in the light of a conflict with duty or with another basic goal). On the other hand it is far from certain that they are coercive in the sense that morality or jurisprudence can regard them as exculpatory. For instance a person's intention to maximise his pleasure cannot without further ado be accepted as an exculpatory one. A considerably more plausible candidate as an exculpatory intention is where the agent wants to free himself of a very strong or even unbearable tension. If psychiatry does indeed think that the sensations experienced by the kleptomaniac or pyromaniac whose drive remains unsatisfied are unbearably uncomfortable or unbearably importunate, then it is reasonable to look on these sensations as morally legitimising, to look upon them as constituting imperative reasons for the action by which the drive is satisfied.

I have attempted thus far to identify the compulsion embodied in the behaviour of the paranoiac or of the person following a drive, but where it nevertheless can be presupposed that the agent has a certain rationality and in principle can be influenced by moral reasoning. A paranoiac can of course have a completely rational hierarchy of goals, and have a very sharp and lively intellect too — the only thing is that his notion of reality is in one specific respect utterly distorted. Apart from his drive, the kleptomaniac or pyromaniac can be similarly "normal".

Consider now a slightly different category, one which is probably rather common in the statistics of criminality. I have in mind the persons with very low intellectual agility and a poor or even non-existent capacity for weighing one goal against another. Normally their goals are quite rudimentary and unlikely to be enriched. How should their criminal behaviour be viewed from our present point of view?

In order to simplify the argument consider a quite extreme case. Assume that a person has the goal of satisfying a basic drive to kill. Assume also that he does not see, and indeed does not even have the capacity to see, any goal, moral or otherwise, which is in conflict with this goal. As a matter of brute fact his goal is then not open to negotiation. Assume that he kills a particular man to satisfy his drive and cannot see any alternative to this behaviour in order to achieve the satisfaction of the drive. His drive to kill is then coercive in our general sense of the word.

But should his behaviour be excused? There seems to be a difference between this case and the previous ones, in the following respect. In the case of the intellectually agile, society must approve of or accept the coercive goals. But here society cannot in any sense accept the goal as legitimate. The primitive man cannot be excused in terms of self-defence or in terms of unbearable pain. The excusing reason must be different.

It seems to me obvious that the excusing reason must simply be his inability to see any alternative behaviour. The primitive man could not have acted otherwise than he did because there was not, and could not have been, any alternative goal available. (Observe that this case can be treated also with the help of the second criterion of accountability inasmuch as the agent does not know (and cannot know) that his behaviour is morally wrong.)

Let us now proceed on the assumption that we have elucidated the difference between the concepts of sufficient condition and compulsion. It is possible that all our actions are sufficiently conditioned, but it is not the case that all our acts are compulsive in the sense just analysed.

Compulsion is also accorded clear significance in the Swedish Penal Code. If a certain type of compulsion is present a perpetrator can be exempted from punishment. From this it is a short step to proposing a concept of accountability which is connected to the concept of compulsion, as follows:

> A person *A* is accountable with respect to the action *F* only if *A* is not compelled to perform *F*.

Of course the compulsion is here to be understood as being related to a restricted class of intentions legitimised by society. A person is to be regarded as compelled to perform a certain action if he considers his life to be threatened, or if there is threat of serious injury to himself or a member of his family, or if he is suffering under an unbearable inner tension or drive, or if he cannot see any alternative goal to the one he has. This gives accountability an empirical content which still has to do with the determination situation. The concept retains a connection to the original one. I have offered a new analysis of the third clause of accountability, which concerns the person who could not have acted otherwise. The new analysis at the same time gives an interpretation of when an action is to be regarded as excused.

But we still have not taken up a position on the reason for making an exception in criminal law in the case of crimes committed by persons who are not accountable. Why should a crime committed under compulsion be exempt from penal sanction? Let us here go into the matter from two separate points of view — firstly that of a contract theory of criminal law, and secondly that of a utilitarian theory.

Assume that we are in the perfect contract-theory situation. We are participating in the formulation of rules for a decent society. We participate in the decision as to what sanctions are to be applied if we or anyone else should violate these rules, and also in the decision as to where exceptions are to be made. Under what circumstances do we think, for instance, that it would be unreasonable that we ourselves should be subjected to sanction if we were caught committing a crime? (Compare this with the situation in civil law. Under what circumstances do we find it unreasonable that a damages clause should come into effect in the case of breach of contract?)

We regard it as unreasonable that we should be punished, or have to pay damages, for an action which we are, in the analysed sense, compelled to perform. It would, for instance, be unreasonable to submit, for the sake of remaining law-abiding, to serious injury or even death. Such exceptions cannot be regarded as a threat to the purpose of the contract; they concern only a small and fairly clearly defined class of acts.

How, then, are we to regard compulsion as exculpatory from a strictly utilitarian point of view? At present, at least, the (utilitarian) Penal Code makes a distinction between external and inner compulsion: the former is exculpatory, the latter is not. Now, is there any acceptable reason for this?

From the individual-preventive point of view there is a clear rational ground for the distinction. A person who commits a crime because of an external threat does not himself need to be reformed in order to refrain from committing the same type of crime in the future, whereas a person who commits a crime because of inner compulsion which derives from illness can in principle be reformed through medical treatment of the illness. In the former case society ought, wherever possible, to intervene in order to change certain situational conditions, and ought to intervene in the latter case in order to change certain intrapersonal conditions. In the former case, therefore, there is no justification for personal sanction, whereas in the latter case there is.

This reasoning may seem clear enough, yet I do not find it wholly satisfactory. It appears to me that even from the individual-preventive point of view a distinction has to be made between different sorts of crime prevention measures. It is altogether inadequate to distinguish only between such measures as are directed at external circumstances and such as are directed at the perpetrator himself. Those directed at the perpetrator himself can be of very different logical character. On the one hand they can be directed at his *person* (that is to say, his moral and cognitive self) and be intended to remind him of the reprehensible nature of the crime, appealing at the same time to his conscience. This is the original and normal influencing function of punishment. On the other hand the measures can be directed at the perpetrator's *body* in an attempt to influence certain basic biological or psychological conditions of his behaviour. This is completely manipulative and has nothing to do with moral education.

So long as the criminal law does not put on a par all crime prevention measures that follow on crime (including e.g. such measures as are aimed at external circumstances or at persons other than the perpetrator), and call them sanctions, it is merely misleading to lump together those that are communicatively aimed at the perpetrator's person and those that are manipulatively aimed at his body (and psyche).

If only such measures as are aimed at the perpetrator are to be called sanctions, there remains an extremely strong similarity between sanction and the concept of punishment in the old Criminal Code. This concept had a strong moral charge: punishment was among other things a means by which society administered justice and expressed moral indignation.

But in fact the purpose of the introduction of the concept of sanction was to get right away from this manner of looking at things. This can scarcely be achieved, however, by letting courts sentence mentally disordered perpetrators to medical treatment (be it in the form of brain surgery or of psychiatric care) in the same way as they sentence mentally healthy perpetrators to imprisonment, and then referring to both these effects of the law as sanctions, whilst other crime prevention measures (not directly aimed at the perpetrators) do not have the status of sanctions. For we then have an inconsistency in the system which has the effect if anything of preserving our traditional notions of punishment and sanction.[58]

---

58 From a utilitarian point of view a possible alternative to the system laid out in the Swedish Penal Code would be the transference of "surrender for special care" from the class of sanctions to that of other legal effects of crime. At present the latter class consists in the law of, for instance, confiscation, deportation and loss of driving licence.

PART IV

Summary of Basic Concepts

GLOSSARY

EPISODES

State: An object x is in a state s at time t, if and only if x at t has a contingent property or stands in a contingent relation to some other object.

Process: x is in a process at time t, if and only if there is a property- or relation-dimension D pertaining to x such that x's position at t with respect to D is distinct from x's position at $t\text{-}1$ with respect to D, and moreover such that x's position at $t\text{-}1$ is distinct from x's position at $t\text{-}2$ with respect to D.

Event: An event such that x comes into the state s or enters the process pr at time t, occurs, if and only if x is in s or pr at t, and it is not true that x was in s or pr at $t\text{-}1$.

Causative episode: An episode involving both a cause and an effect, as in "the gale broke the window". Causative episodes may be classified as causative states, processes and events, depending on the nature of the entailed effect-episode.

Agentive episode (action): An episode involving both an agent *A* and the result brought about by the agent.

A performs an action, if and only if *A* intentionally makes something be the case or lets something be the case in the world. Agentive episodes may be classified as agentive states, agentive processes (i.e. activities) or agentive events (i.e. acts). Most actions are at the same time causative episodes. When *A* opens the door *A* causes it to be the case that the door is open.

Omission: An action which is not a causative episode. *A* omits to bring about *P*, if and only if *A* intentionally lets *non-P* remain the case.

Intention: *A* intends to bring about *P*, only if *A* is in a state which is such that

(i) *A* believes that she can bring about *P*, and
(ii) For all actions *X*, if *A* believes that she cannot bring about *P* unless she does *X*, and if *A* can do *X*, then *A* will do *X*.

Decision: *A* decides to bring about *P* means that *A* forms the intention to bring about *P*.

Want: *A* wants to bring about *P*, if and only if *A* is in a state which is such that, if *A* believes that she can bring about *P* and if there is no overriding factor, such as a conflicting intention, then *A* will form the intention to bring about *P*.

COMPLEX EPISODES

Episode embedment (the causative case): the object o causes an episode E by causing another episode $E1$. Here $E1$ is said to be *embedded* in E (or collateral to E).

Episode embedment (the agentive case): the agent *A* does *F* by doing *F1*. Here *F1* is embedded in *F*.

Basic action: *F* is a basic action, if and only if *F* does not embed any action distinct from *F* itself.

Generated action: An action *F* which embeds another action *F1* is said to be generated by *F1*. *F1* may be a basic action or itself a generated action. The relation of action-generation is transitive.

Accomplishment: *F* is an accomplishment, if and only if *F* is a generated action which embeds only one basic action.

Project: *P* is a project, if and only if *P* is a generated action which embeds a sequence entailing at least two basic actions.

Some Other Concepts

Intentional explanation of actions (schema for):
Explanandum: *A* does *F*
Explanans:

(i) *A* intends to bring about *P*
(ii) *A* believes that she is in situation *S*
(iii) *A* believes that in *S* she will not bring about *P* unless she does *F*
(iv) *A* can do *F*

Practical possibility (of performing an action): *A* has the practical possibility of doing *F*, if and only if *A* is both able and has the opportunity to do *F*.

First-order ability: *A* is able to do *F*, relative to an opportunity *O*, if and only if *A* is in a state which is such that, if *A* tries to do *F* in *O*, then *A* does *F*.

Second-order ability: *A* has the second-order ability to do *F*, if and only if *A* has the first-order ability, relative to an opportunity *O*, to pursue a training programme after the completion of which *A* will have the first order ability to do *F*.

Ability (in a health context): *A* is able to do *F*, relative to an accepted set of circumstances *C*, if and only if *A* is in a state which is such that, if *A* tries to do *F* in *C*, then *A* does *F*.

Health (the general notion): *A* is completely healthy, if and only if *A* is in a bodily and mental state such that *A* has the second-order ability, given an accepted set of circumstances *C*, to realise all her vital goals.

Illness: *A* is to a certain extent ill, if and only if there is some vital goal of *A*'s such that *A* cannot fully realise it, given an accepted set of circumstances.

Disease: A bodily or mental process which is such that it tends to reduce the health of its bearer. (Similar definitions can be given for *injury* and *defect* by substituting "acquired state" and "congenital state" for "process" in the definition of disease.)

Health (the medical notion): *A* is medically healthy, if and only if *A*'s health (in the general sense) is not compromised by any of the malady-forms: disease, injury or defect.

Vital goals: The vital goals of *A* are the set of states which are such that their realisation is necessary and jointly sufficient for *A*'s minimal long-term happiness.

Happiness: *A* is completely happy at *t*, if and only if

(i) *A* wants at *t* that (*x1...xn*) shall be the case at *t*
(ii) (*x1...xn*) constitutes the totality of *A*'s wants at *t*
(iii) *A* finds at *t* that (*x1...xn*) is the case.

BIBLIOGRAPHY

[1] Agich, G.J.: 1983, 'Disease and Value: A Rejection of the Value-neutrality Thesis', *Theoretical Medicine, 4*, 27-41.

[2] Aristotle: 1934, *The Nicomachean Ethics*, The Loeb Classical Library, Harvard University Press, Cambridge, Massachusetts.

[3] Aristotle: 1934, *Physics*, *Books V-VIII,* The Loeb Classical Library, Harvard University Press, Cambridge, Massachusetts.

[4] Aristotle: 1982, *The Eudemian Ethics*, *Books I, II, and VIII*, translated with a Commentary by M. Woods, Clarendon Aristotle Series, Clarendon Press, Oxford.

[5] Audi, R.: 1997, 'Intending and its Place in the Theory of Action', in Holmström-Hintikka, G. and Tuomela, R. (eds.): *Contemporary Action Theory, vol. I: Individual Action,* Kluwer Academic Publishers, Dordrecht, pp. 177-196.

[6] Ayers, R. R.: 1968, *The Refutation of Determinism*, Methuen & Co, London.

[7] Badley, E. M.: 1987, 'The ICIDH: format, application in different settings, and distinction between disability and handicaps', *International Disability Studies*, *9*, 118-122.

[8] Beauchamp, T. L. and Childress, J.F.: 1989, *Principles of Biomedical Ethics*, Oxford University Press, Oxford.

[9] Boorse, C.: 1975, 'On the Distinction between Disease and Illness', *Philosophy and Public Affairs*, *5*, 49-68.

[10] Boorse, C.: 1976, 'What a Theory of Mental Health Should Be', *Journal for the Theory of Social Behavior*, *6*, 61-84.

[11] Boorse, C.: 1977, 'Health as a Theoretical Concept', *Philosophy of Science*, *44*, 542-573.

[12] Boorse, C.: 1997, 'A Rebuttal on Health', in Humber, J. and Almeder, R. (eds.): *What is Disease?*, Biomedical Ethics Reviews, Humana Press, New Jersey.

[13] Braybrooke, D.: 1987, *Meeting Needs: Studies in Moral, Political and Legal Philosophy*, Princeton University Press, Princeton.

[14] Broad, C.D.: 1954, 'Emotion and Sentiment', *Journal of Aesthetics and Art Criticism*, *12*, 203-214.

[15] Bury, M. R.: 1987, 'The ICIDH: a review of research and prospects', *International Disability Studies*, *9,* 122-125.

[16] Canadian Society for the ICIDH: 1989, 'Consultation: Proposal for the Revision of the Third Level of the ICIDH: the Handicap', *ICIDH International Network*, *vol. 2,* number 1.

[17] Canadian Society for the ICIDH: 1990, 'Comments on the Proposition of Revision of the Handicap Concept', *ICIDH International Network*, *vol. 3*, number 2.

[18] Canguilhem, G.: 1978, *On the Normal and the Pathological*, D. Reidel Publishing Company, Dordrecht.

[19] Churchland, P.: 1970, The 'Logical Character of Action Explanations', *Philosophical Review*, *79*, 214-236.

[20] Cohen, G.A.: 1993, 'Equality of What? On Welfare, Goods and Capabilities', in Nussbaum, M. and Sen, A. (eds)., *Quality of Life*, Clarendon Press, Oxford, pp. 9-29

[21] Culver, C.M. and Gert, E.: 1982, *Philosophy in Medicine: Conceptual and Ethical Issues in Medicine and Psychiatry*, Oxford University Press, Oxford.

[22] Davidson, D.: 1980, *Essays on Actions and Events*, Clarendon Press, Oxford.

[23] D'Arcy, E.: 1966, *Human Acts: An Essay in Their Moral Evaluation*, Clarendon Press, Oxford.

[24] Downie, R. S., Fyfe, C. and Tannahill, A.: 1990, *Health Promotion: Models and Values*, Oxford University Press, Oxford.

[25] Dray, W.H.: 1957, *Laws and Explanation in History*, Oxford University Press, Oxford.

[26] Dworkin, G.: 1988, *The Theory and Practice of Autonomy*, Cambridge Studies in Philosophy, Cambridge University Press, Cambridge.

[27] Edwards, S.D.: 1997, 'Dismantling the Disability/Handicap Distinction', *The Journal of Medicine and Philosophy*, *22*, 589-606.

[28] Engelhardt, H.T. Jr.: 1995, *The Foundations of Bioethics*, Second, revised, edition, Oxford University Press, Oxford.

[29] Faden, R.R. and Beauchamp, T.L.: 1986, *A History and Theory of Informed Consent*, Oxford University Press, Oxford.

[30] Fulford, K.W.M.: 1989, *Moral Theory and Medical Practice*, Cambridge University Press, Cambridge.

[31] Fulford, K.W.M.: 1990, 'Philosophy and Medicine: The Oxford Connection', *British Journal of Psychiatry*, *157*, 111-115.

[32] Fulford, K.W.M.: 1993, 'Praxis Makes Perfect: Illness as a Bridge Between Biological Concepts of Disease and Social Conceptions of Health', *Theoretical Medicine*, *14*, 305-320.

[33] Goldman, A.I.: 1970, *A Theory of Human Action*, Prentice Hall Inc., New Jersey.

[34] Green, O.H.: 1993, *The Emotions: A Philosophical Theory*, Kluwer Academic Publishers, Dordrect.

[35] Griffin, J.: 1986, *Wellbeing: Its Meaning, Measurement, and Moral Importance*, Clarendon Press, Oxford.

[36] Hempel, C.G.: 1965, *Aspects of Scientific Explanation and Other Essays in the Philosophy of Science*, The Free Press, New York.

[37] Hilpinen, R.: 1997, 'On States, Actions, Omissions and Norms', in Holmström-Hintikka, G. and Tuomela, R. (eds.): *Contemporary Action Theory, vol.1: Individual Action*, Kluwer Academic Publishers, Dordrecht, pp.83-107.

[38] Holmström-Hintikka, G. and Tuomela, R. (eds.): 1997, *Contemporary Action Theory, vol.1: Individual Action*, Kluwer Academic Publishers, Dordrecht.

[39] Hook, S. (ed.): 1963, *Philosophy and History,* New York University Press, New York.

[40] *ICD (Manual of the International Classification of Diseases and Related Health Problems)*: 1992, 10th Revision, Publication of the World Health Organisation, Geneva.

[41] *ICIDH* (*International Classification of Impairments, Disabilities and Handicaps*): 1993, Publication of the World Health Organisation, Geneva.

[42] Kanger, S.: 1972, 'Law and Logic', *Theoria, XXVIII,* 105-132 .

[43] Kenny, A.: 1963, *Action, Emotion, and Will*, Routledge & Kegan Paul, London.

[44] Kenny, A.: 1975, *Will, Freedom and Power*, Basil Blackwell, Oxford.

[45] Kinberg, O.: 1930, *Aktuella kriminalitetsproblem i psykologisk belysning* (Current Problems of Criminality in Psychological Light), Stockholm.

[46] Kóvacs. J.: 'The Concept of Health and Disease', *Medicine, Health Care and Philosophy, 1,* 31-39.

[47] Leontiev, A.N.: 1981, 'The Problem of Activity in Psychology', in Wertsch J.W. (ed.): *The Concept of Activity in Soviet Psychology*, M.E. Skarpe, Ar'monk, New York, pp. 37-71.

[48] Liss, P-E.: 1993, *Health Care Need*, Avebury, Aldershot.

[49] Liss, P-E. and Nikku, N. (eds.): 1994, *Health Promotion and Prevention: Theoretical and Ethical Aspects*, Swedish Council for Planning and Coordination of Research, Report 94:4.

[50] Mackie, J.L.: 1974, *The Cement of the Universe*, Oxford University Press, Oxford.

[51] Marc-Wogau, K.: 1967, 'On Historical Explanation', in *Philosophical Essays*, Library of Theoria, *NO XI*, pp. 223-240.

[52] Marinker, J.H.: 1975, 'Why Make People Patients?', *Journal of Medical Ethics*, *I*, 81-84.

[53] Marras, A.: 1997, 'Metaphysical Foundations of Action Explanation', in Holmström-Hintikka, G. and Tuomela, R. (eds.): *Contemporary Action Theory, vol.1: Individual Action*, Kluwer Academic Publishers, Dordrecht, pp. .

[54] Maslow, A.: 1954, *Motivation and Personality*, Harper & Row, New York.

[55] Maslow, A.: 1968, *Toward a Psychology of Being*, D. van Nostrand Company, New York.

[56] Melden, A.I.: 1961, *Free Action*, Routledge & Kegan Paul, London.

[57] Mele, R.M. (ed.): 1997, *The Philosophy of Action*, Oxford University Press, Oxford.

[58] Menninger, K.: 1968, *The Crime of Punishment*, Penguin, New York.

[59] Morriss, P.: 1987, *Power: A Philosophical Analysis*, S:t Martin's Press, New York.

[60] Moya, C.J.: 1990, *The Philosophy of Action: An Introduction*, Basil Blackwell, Oxford.

[61] National Council for Crime Prevention: 1990, *The Swedish Penal Code*, 1990:3, Stockholm.

[62] Nelson, A.: 1970, *Brott och påföljd: En orientering om svensk straffrätt enligt brottsbalken* (Crime and its Legal Consequences: An Introduction to Swedish Penal Statutes According to the Penal Code), Prisma, Stockholm.
[63] Nordenfelt, L.: 1974, *Explanation of Human Actions*, Dept of Philosophy and the Philosophical Society, Uppsala University, Uppsala.
[64] Nordenfelt, L.: 1977, *Events, Actions and Ordinary Language*, Doxa, Lund.
[65] Nordenfelt, L.: 1981, *Causation: An Essay*, Akademilitteratur, Stockholm.
[66] Nordenfelt, L.: 1983/1997, *On Disabilities and Their Classification*, Studies on Health and Society, SHS *1*, University of Linköping, Linköping.
[67] Nordenfelt, L.: 1991, *Towards a Theory of Health Promotion*, Health Service Studies *5*, Linköping.
[68] Nordenfelt, L.: 1992, *On Crime, Punishment, and Psychiatric Care*, Almqvist & Wiksell International, Stockholm.
[69] Nordenfelt, L.: 1993, *Quality of Life, Health and Happiness*, Avebury, Aldershot.
[70] Nordenfelt, L. (ed.): 1994, *Concepts and Measurement of Quality of Life in Health Care*, Kluwer Academic Publishers, Dordrecht.
[71] Nordenfelt, L.: 1995 a, *On the Nature of Health*, Second, revised, edition, Kluwer Academic Publishers, Dordrecht.
[72] Nordenfelt , L. 1995 b, 'On the General Concepts of Action and Ability', *International Journal of Technology Assessment in Health Care*, *11*, 145-152.
[73] Nordenfelt, L.: 1997 a, 'On Ability, Opportunity and Competence: An Inquiry into People's Possibility for Action', in Holmström-Hintikka,G. and Tuomela, R. (eds.): *Contemporary Action Theory, vol 1: Individual Action*, Kluwer Academic Publishers, Dordrecht, pp. 145-158.
[74] Nordenfelt, L.: 1997 b, 'The Importance of a Disability/Handicap Distinction', *The Journal of Medicine and Philosophy*, *22*, 607-622.
[75] Nordenfelt, L.: 1998, 'On Medicine and Other Means of Health Enhancement', *Medicine, Health Care and Philosophy*, *1*, 1-12.
[76] Parsons, T.: 1972, 'Definitions of Health and Illness in the Light of American Values', in Jaco, E.G. (ed.): *Patients, Physicians and Illness*, The Free Press, New York, pp.107-127.
[77] Pörn, I.: 1970, *The Logic of Power*, Basil Blackwell, Oxford.
[78] Pörn, I.: 1977, *Action Theory and Social Science*, D. Reidel Publishing Company, Dordrecht.
[79] Pörn, I.: 1984, 'An Equilibrium Model of Health', in Nordenfelt, L. and Lindahl, B.I.B. (eds.): *Health, Disease, and Causal Explanations in Medicine*, D. Reidel Publishing Company, Dordrecht.
[80] Pörn, I.: 1986, 'On the Nature of Emotions', in Needham, P. and Odelstad, J. (eds.): *Changing Positions: Essays Dedicated to Lars Lindahl*, Philosophical Studies, University of Uppsala, Uppsala, pp. 205-214
[81] Pörn, I.: 1993, 'Health and Adaptedness', *Theoretical Medicine*, *14*, 295-303.
[82] Radden, J.: 1985, *Madness and Reason*, George Allen & Unwin, London.
[83] Rawls, J.: 1972, *A Theory of Justice*, Oxford University Press, Oxford.
[84] Rundle, B.: 1997, *Mind in Action*, Clarendon Press, Oxford.

[85] Salmon, W.C.: 1984, *Scientific Explanation and the Causal Structure of* the *World*, Princeton University Press, Princeton.
[86] Sandøe, P. and Kappel, K.: 1994, 'Changing Preferences: Conceptual Problems in Comparing Health-Related Quality of Life', in Nordenfelt, L. (ed.): *Concepts and Measurement of Quality of Life in Health Care*, Kluwer Academic Publishers, Dordrecht, pp. 161-180.
[87] Seedhouse, D.: 1986, *Health: The Foundations of Achievement*, John Wiley & Sons, Chichester.
[88] Seedhouse, D.: 1988, *Ethics: The Heart of Health Care*, John Wiley & Sons, Chichester.
[89] Seedhouse, D.: 1991, *Liberating Medicine*, John Wiley & Sons, Chichester.
[90] Searle, J.R.: 1983, *Intentionality: An Essay in the Philosophy of Mind,* Cambridge University Press, Cambridge.
[91] Sen, A.: 1985, *Commodities and Capabilities*, North-Holland, Amsterdam.
[92] Sen, A.: 1992, *Inequality Reexamined*, Clarendon Press, Oxford.
[93] Sen A.: 1993, 'Capability and Well-being', in Nussbaum, M. and Sen, A. (eds.): *Quality of Life*, Clarendon Press, Oxford, pp. 30-49.
[94] Shwayder, D.S.: 1965, *The Stratification of Behaviour*, Routledge & Kegan Paul, London.
[95] Sundström, P.: 1987, *Icons of Disease*, Linköping Studies in Arts and Science, *14*, Linköping University, Linköping.
[96] Talja, J.: 1985, 'On the Logic of Omissions', *Synthese*, *65*, 235-248.
[97] Tatarkiewiz, W.: 1976, *Analysis of Happiness*, Martinus Nijhoff, The Hague.
[98] Taylor, C.: 1985, *Human Agency and Language*, *Philosophical Papers I,* Cambridge University Press, Cambridge.
[99] Telfer, E.: 1980, *Happiness*, Macmillan, London.
[100] Ten, C.L.: 1987, *Crime, Guilt and Punishment*, Oxford University Press, Oxford.
[101] Temkin, O.: 1963, 'The Scientific Approach to Disease: Specific Entity and Individual Sickness', in Crombie, A.C. (ed.): *Scientific Change: Historical Studies in the Intellectual, Social and Technical Conditions for Scientific Discovery and Technical Invention from Antiquity to the Present*, Basic Books, Inc., New York, pp. 629-647.
[102] Tuomela, R.: 1977, *Human Action and Its Explanation*, D. Reidel Publishing Company, Dordrecht.
[103] Tuomela, R.: 1982, 'Explanation of action', in Fløistad, G. (ed.): *Contemporary Philosophy*, *vol. 3*, Martinus Nijhoff Publishers, The Hague, pp. 15-43.
[104] Tuomela, R.: 1995, *The Importance of Us: A Philosophical Study of Basic Social Notions*, Stanford University Press, Stanford.
[105] Urmson, J.O.: 1988, *Aristotle's Ethics*, Oxford University Press, Oxford.
[106] Van Fraassen, B.C.: 1980, *The Scientific Image*, Clarendon Paperbacks, Oxford.
[107] Vendler, Z.: 1967, *Linguistics in Philosophy*, Cornell University Press, Ithaka.
[108] Wertheimer, A.: 1987, *Coercion*, Princeton Unviersity Press, Princeton.

[109] Whitbeck, C.: 1981, 'A Theory of Health', in Caplan, A.L., Engelhardt, H.T. Jr., and McCartney, J.J. (eds.): *Concepts of Health and Disease: Interdisciplinary Perspectives*, Addison-Wesley Publishing Company, Reading, Massachusetts, pp. 611- 626.

[110] Wittgenstein, L.: 1953, *Philosophical Investigations*, Basil Blackwell, Oxford.

[111] von Wright, G.H.: 1963, *The Varieties of Goodness*, Routledge & Kegan Paul, London.

[112] von Wright, G.H.: 1963, *Norm and Action,* Routledge & Kegan Paul, London.

[113] von Wright, G.H.: 1968, *An Essay in Deontic Logic and the General Theory of Action*, Acta Philosophica Fennica, North Holland, Amsterdam..

[114] von Wright, G.H: 1971, *Explanation and Understanding*, Routledge & Kegan Paul, London.

[115] von Wright, G.H.: 1975, 'Determinism and the Study of Man', in Manninen, J. and Tuomela, R. (eds.): *Essays on Explanation and Understanding*, D. Reidel Publishing Company, Dordrecht, pp. 415-435.

[116] von Wright,G.H.: 1997, 'Explanation and Understanding of Actions', in Holmström-Hintikka, G. and Tuomela, R. (eds.): *Contemporary Action Theory, vol 1: Individual Action*, Kluwer Academic Publishers, Dordrecht, 1-20.

INDEX

MIX
Papier aus verantwortungsvollen Quellen
Paper from responsible sources
FSC® C105338

If you have any concerns about our products,
you can contact us on
ProductSafety@springernature.com

In case Publisher is established outside the EU,
the EU authorized representative is:
Springer Nature Customer Service Center GmbH
Europaplatz 3, 69115 Heidelberg, Germany

Printed by Libri Plureos GmbH
in Hamburg, Germany